Preface

One of the many stories told of Ernest Rutherford is his reply when he was congratulated upon a particularly creative discovery. "You are a lucky man, Rutherford, you are always on the crest of a wave." To which Rutherford observed: "Well, I made the wave, didn't I?" Then, after a short pause, he added, "At least to some extent". The moral of this anecdote is that even the most creative of scientific thoughts or experiments are dependent upon the work of earlier investigators.

Of the writings of histories of chemistry there can be no end. In the 1980s and 1990s there was a flourish of such historical activity. Three histories of chemistry were published in English in 1991, three more (one of them in French) in 1992. One of these contributions was David Knight's *Ideas in Chemistry: A History of the Science* (1992), which was not a history of chemistry in the conventional sense. It did not offer the reader a chronological or thematically detailed account of the development of chemistry since antiquity. Rather, it was an account of scientific idea and practices in the form of self-contained essays in which chemistry was used as the exemplar. For Knight, chemistry was a "finished science". He did not mean by this that, in future, chemistry would spring no surprises (Buckminster fullerenes were, after all, the chemical sensation of the 1990s), but that with its reduction to the behaviour of sub-atomic particles and electron clouds, chemistry was no longer at the leading edge of scientific ideas but instead,

The Case of the Poisonous Socks: Tales from Chemistry
By William H. Brock
© William H. Brock, 2011
Published by the Royal Society of Chemistry, www.rsc.org

par excellence, the central service science which underwrites our understanding of biology, medicine, solid state physics and the whole gamut of everyday technology. In this sense, it is an ideal "dead science" whose biography can be written, rather like that of a writer or artist whose career and work is complete but whose paradigmatic inspiration lives on and whose writings or canvases invoke a plethora of normal science puzzle-making and puzzle-solving.

The essay form has had a long tradition in the history of chemistry, beginning in the seventeenth century with Robert Boyle's effusive reports on colours and many other subjects and proceeding, notably, through Richard Watson's *Chemical Essays* (1781) and Justus Liebig's *Familiar Letters on Chemistry* (1843) to the twentieth century, with the thought-provoking writings of Primo Levi and Erwin Chargaff.

Liebig's essays, which began as informative newspaper articles, grew in book publication into vehicles for arguing a point of view and for paying off old scores, but above all for demonstrating that chemistry was essential knowledge for understanding plant and human physiology and for solving problems of health and nutrition, as well as saving the world from Malthusian agricultural damnation and misery. Inspired by Liebig's example, from the end of the nineteenth century until the 1940s it was common for historically-minded chemists to publish collections of their occasional contributions to non-scientific periodicals, informal talks and speeches to clubs and societies, and obituaries of chemical colleagues with whom they had trained or collaborated. Many of these, like August Wilhelm Hofmann's *Zur Erinnerungen an vorangegangene Freunde* (1888), W. A. Shenstone's *The New Physics and Chemistry* (1906), James Campbell Brown's *Essays and Addresses* (1914), Frederick Soddy's *Science and Life* (1920) and John Read's *Through Alchemy to Chemistry. A Procession of Ideas and Personalities* (1957) are still consulted by historians today.

The present book contains essays and reviews written over the last forty years after I left chemistry to practise as an historian of chemistry. The essays have been revised and updated and shorn of academic documentation to make their reading simpler and, I hope, more enjoyable for historically-minded chemists, as well as for general readers. The collection is divided into six sections, each with its own introduction. The first, which is arranged in a rough chronological order, contains chemical tales which explore some of the ways that chemistry has seen its future. The second looks at the social history of chemistry and the way it has been organized. A third section offers vignettes of some major figures in chemistry. The fourth shows that women chemists have had a significant role in the development of chemistry. A fifth group of essays deals light-heartedly with a number of chemistry books and journals that have intrigued me. The book ends by examining a number of chemists who, for various reasons, have made their names in different fields. Some suggestions for further reading are given for readers who would like to follow up any of my chemical tales.

"Books like this", wrote Knight in his *Ideas in Chemistry*, "resemble paintings done with a broad brush, and are successful if they stimulate readers to take an interest in the history of chemistry". Like Liebig, my aim in this book is to beget a better public understanding of science, *qua* chemistry, *via* an historical viewpoint. It is in tune with the feelings of most responsible historians of science that our subject has a role to play in public and scientific education. Because science is the most important civilizing influence since Christendom, our historical interpretations deserve a wider audience than that of the seemingly diminishing number of professional historians of science.

Contents

Part 1: Chemical Futures

Chapter 1
The Case of the Poisonous Socks   3

Chapter 2
Taste, Smell and Flavour   9

Chapter 3
Tales of Hofmann   16

Chapter 4
Liebig on Toast   20

Chapter 5
**The Future of Research at the Royal Institution (London) and
the Smithsonian Institution (Washington)**   24

Chapter 6
The Future of Chemistry in 1901   40

Chapter 7
The Alchemical Society, 1912–1915   48

The Case of the Poisonous Socks: Tales from Chemistry
By William H. Brock
© William H. Brock, 2011
Published by the Royal Society of Chemistry, www.rsc.org

Part 2: Organizing Chemistry

Chapter 8
Putting the 'S' in the 'Three R's'   57

Chapter 9
The London Chemical Society, 1824   67

Chapter 10
The State of Chemistry in Britain in 1846   74

Chapter 11
The Laboratory Before and After Liebig   85

Chapter 12
The Chemical Origins of Practical Physics   97

Chapter 13
Chemical Algebra   114

Chapter 14
The B-Club   118

Chapter 15
Chemistry by Discovery in a Phrase   124

Part 3: A Cluster of Chemists

Chapter 16
Amedeo Avogadro   131

Chapter 17
**Liebig and Wöhler: Creating a Path through the Dark Province
of Organic Nature**   138

Chapter 18
August Kekulé (1829–1896): Theoretical Chemist   147

Chapter 19
**The Don Quixote of Chemistry: Sir Benjamin Collins Brodie
(1817–1880)**   158

Chapter 20
The Epistle of Henry the Chemist 166

Chapter 21
He Knew He Was Right—Fritz Haber 176

Chapter 22
J. R. Partington (1886–1965): Physical Chemistry in Deed and Word 180

Chapter 23
Henry Crookes, Founder of Crookes Laboratories 192

Chapter 24
A Life of Magic Chemistry 202

Part 4: Women Chemists

Chapter 25
Women in Alchemy and Chemistry 207

Chapter 26
Teaching Chemistry to Women 211

Chapter 27
Musical Affinities 214

Chapter 28
Edith Hilda Usherwood (1898–1988) and the Ingold Partnership 218

Part 5: Chemical Books and Journals

Chapter 29
The Fate of Eponymous Chemical Journals 233

Chapter 30
The Lamp of Learning: Richard Taylor and the Textbook 237

Chapter 31
"The Greatest Work which England has ever Produced": Henry Watts and the *Dictionary of Chemistry* 244

Chapter 32
Chemistry in the Aquarium   **248**

Chapter 33
Insurance Chemistry   **255**

Chapter 34
Math for Chemists   **262**

Chapter 35
The Chemistry of Pottery   **267**

Chapter 36
Baker's Dozens   **271**

Part 6: Lost to Chemistry

Chapter 37
They Also Ran   **279**

Chapter 38
Who Was Crookes's Musician–Chemist?   **282**

Chapter 39
The Chemist from Hanwell Asylum   **286**

Chapter 40
George Du Maurier (1834–1896)   **290**

Chapter 41
Sir Stafford Cripps   **294**

Chapter 42
The Chemist as Novelist: The Case of C. P. Snow   **301**

Further Reading   **314**

Subject Index   **327**

Sources and Acknowledgements

For permission to publish revised versions of essays that appeared originally in their publications I am grateful to: the Historical Group of the Royal Society of Chemistry and editor of its newsletter (1, 14, 38, 40, 41); the Wellcome Trust and editor of *Medical History* (2); the Society of Chemical Industry and editor of *Chemistry & Industry* (4, 21, 24, 27); Rodopi Publishers, Amsterdam (5); the Society for the History of Alchemy and Chemistry's *Ambix* (9, 10); Excelsior Publications, Paris (11); the American Chemical Society's Historical Division and editor of *Bulletin for the History of Chemistry* (12, 22); the chairman of Prinz Albert Gesellschaft, Coburg (8); the editor of *CHEM13 News*, University of Waterloo (15); the editors of the Royal Society of Chemistry's *Education in Chemistry* and *Chemistry in Britain* (16, 20, 42); Elsevier and the editor of *Endeavour* (18); the editor of *Hyle* (19); the editor of *Archives of Natural History* (32); the Chemical Heritage Foundation, Philadelphia (29, 31, 33–36); and the editor of *Chem@Cam* (37). For allowing me access to archives in their care, I am grateful to Jane May (Leicestershire Museums, 1); Mrs Irena McCabe (Royal Institution, 5); William Cox (Smithsonian Institution Archive, 5); David Allen (RSC, 14); Mrs Ann Barrett (Imperial College London, 20); Lorraine Scheene (Queen Mary College, 22); Nichola Court (Royal Society, 22); Anne Thomson (Newnham College, Cambridge, 26); Margaret Harcourt Williams (Rothamsted Research,

The Case of the Poisonous Socks: Tales from Chemistry
By William H. Brock
© William H. Brock, 2011
Published by the Royal Society of Chemistry, www.rsc.org

32); and Margaret Maclean (University of Leicester, 42). My thanks are also due to various individuals for help and suggestions over several years. They include Ted Benfey (formerly Chemical Heritage Foundation), Sabyia Farooq (Leicester), Yoshiyuki Kikuchi (MIT), Peter Morris (Science Museum London), Sir John Rowlinson (Oxford), Andrea Sella (UCL), Anthony S. Travis (Jerusalem) and David Wykes (Dr Williams's Library, London).

Part 1: Chemical Futures

Following previous years in which the work of astronomers and physicists had been celebrated, the United Nations designated 2011 as the International Year of Chemistry (IYC). During the year the achievements of chemists and chemistry's contributions to the well-being of human society were celebrated all over the world. The principal theme chosen for highlight was "chemistry: our life, our future", and this theme provides the underlying thread for the seven varied stories that make up the first section of this book. They bear testimony to the promise—first adumbrated by the alchemists—that chemistry is *the* science that can deliver a brighter and better future for mankind. On the way, however, there can be stumbling blocks and risks which chemists and society itself have to monitor and negotiate. William Crookes, who appears in the first story as a detective searching for the cause of skin irritation caused by the recent introduction of vividly-hued synthetic dyestuffs into the hosiery industry, was a journalist as well as a productive research chemist. He repeatedly used his weekly journal *Chemical News* to publicise and extol the fact that chemistry was improving the human condition. This optimism for a future enriched by chemists culminated shortly after his death in 1919 with Du Pont's catching advertising slogan, "Better Things for Better Living Through Chemistry". This slogan had been no better demonstrated than by Crookes's teacher, the Anglo-German chemist August Wilhelm Hofmann, the subject of the third essay. His transformation of coal tar waste products into beautiful colours exemplifies all that is wonderful and powerful about chemistry.

Although Crookes's predecessor, the London medical chemist William Prout, was more interested in the theoretical basis of our chemical senses, he too looked forward to a time when chemists would understand the principles of nutrition sufficiently to prepare foodstuffs that would transform what and how we eat and drink—as the research of the Japanese chemist Ikeda was to show. Indeed, in later studies, it was Prout who proposed the division of food-stuffs into carbohydrates, fats and proteins, and who developed one of the first biochemical models of the digestive process. The close connection between the chemistry of food and commerce is highlighted in the example of the great nineteenth-century organic chemist, Justus von Liebig, whose work in applied chemistry helped create the modern convenience food industry.

Using chemistry to create a better future is dependent on the funding of fundamental, as well as, applied research. We usually think of such research as being provided by the state, or by industrial or pharmaceutical companies. But as the curious story of George Hodgkins's bequests to the Royal Institution and the Smithsonian Institution in the 1890s reveals, private philanthropy has also had a significant, and sometimes unexpected, role to play. As the individual sciences emerged in their modern form in the nineteenth century, it was inevitable that chemists should have pondered what the future held for their science in the next century and what particular field of chemistry—inorganic, organic or physical—would prove of most value educationally and for research. Although the general view in 1900–1901 was that physical chemistry was the front runner (as turned out to be the case), spokesmen also emphasized the role that history should play in the future education of chemists and the practice of chemistry. This rather surprising conclusion was also underlined in the first part of the twentieth century by the increasing interest in the history of alchemy among both scientists and laymen as the new physics and chemistry of radioactivity revealed the deep structure of the atom. The formation in 1912 of an Alchemical Society, which was treated by *Nature* like any other *bona fide* scientific society, forms the final essay in this section. In the chemistry of the future, chemists were to hand over the atomic nucleus to physicists while retaining the electron as their particular province for enhancing and trans-forming human lives and futures.

CHAPTER 1

The Case of the Poisonous Socks

On 30 September 1868 *The Times* published a police report that
poisons contained in dyes were affecting the public's health. A
certain Dr Webber had complained to a City of London magistrate that brilliantly coloured socks which were being sold locally
had caused severe "constitutional and local complaint" to several
of his patients. In one case, a patient's foot had become so swollen after wearing such socks that his boots had had to be cut off.
Since Webber did not know the nature of the poison, he had
been unable to recommend an antidote, though he found the
application of glycerine gave relief. Red and light brown socks
striped transversely with bright orange and bordered in violet
and black were shown to the magistrate. It seemed that it was
the orange dye that caused the intense irritation. Webber had
made some investigations and found that the orange dye was
made from a new acid and that the dye workers were unable to
work on the substance for more than six months at a time and
that when they "retired" their arms were covered with sores.
Webber had complained to the sock maker, who immediately
stopped an export order for 6,000 pairs of socks, and returned to
using "old wood dyes".

As this reference to natural dyestuffs reminds us, there was
nothing new about coloured socks and stockings. Naturally dyed
silk or worsted hose had been available to the wealthy since the
seventeenth century, while white, grey and black stocking and hose

The Case of the Poisonous Socks: Tales from Chemistry
By William H. Brock
© William H. Brock, 2011
Published by the Royal Society of Chemistry, www.rsc.org

were worn by the majority. Embroidered and coloured "clock goves" (the triangular piece let into the ankle of a sock or stocking) were bought by fashionable gentile society throughout the eighteenth century. Blue dye, produced by using imported indigo bulked out with native woad, was used in the stockings made fashionable by members of the Society of Literati. It was for that reason that satirists described its female members as "blue stockings". By the mid-1850s, dynastic hosiers such as John Morley and Nathaniel Corah seized on the opportunities presented by the new colourful dyes synthesized from aniline to produce the hosiery that Webber complained about.

The magistrate revealed that he was partial to coloured socks himself and had suffered no harm from wearing them. Although urged not to alarm the public unnecessarily, Webber proceeded to reveal that purchasers of the socks in Oxford and Cambridge were also complaining of skin sores. The trouble had even spread to Paris, where English hosiery was on sale. *The Lancet*, in taking up the story, recalled that the previous year a dancer at the Drury Lane Theatre had bought a "gorgeous" pair of socks and ended up in hospital with unpleasant eruptions. The medical journal rightly deduced that the poisoning had something to do with hot sweaty feet causing a chemical reaction with an unknown dyestuff. Obviously, beneath their drab black trousers and skirts, Victorian gentlemen and gentlewomen were secretly cross-gartered Malvolios of fashion!

A few days later, a knowledgeable Coventry physician named McVeigh reported cases of severe eczema (resembling erysipelas) he had seen in the Midlands and blamed aniline dyes retailed as "Marquis of Hastings" colours. The usual symptoms were itchy, swollen, painful, red-hot and blistered feet which also discharged. McVeigh had consulted a recently translated German treatise by Max Reimann, *On Aniline and its Derivatives. A Treatise upon the Manufacture of Aniline and its Colors* (1868), from which he deduced that picric acid was the culprit. This was the cue for the translator of Reimann's book to take up the correspondence. This was none other than the 36-year old chemist, William Crookes who, despite achieving a Fellowship of the Royal Society and an international reputation for the discovery of thallium in 1861, was still struggling to establish a career as chemical consultant and independent researcher. By 1868 he was engaged in a mixture of

Figure 1.1   Sir William Crookes (1832–1919) portrayed in a *Vanity Fair* cartoon at the height of his fame in 1905. (Wellcome Images)

activities ranging from a long project to determine the atomic weight of thallium with great accuracy, the exploitation of patents for using sodium amalgam in silver and gold mining and a carbolic disinfectant, as well as literary activities that ranged from editing the weekly *Chemical News* and the *Quarterly Journal of Science* to making translations of German technical works. As Reimann's translator, he seized upon the controversy to demonstrate his expertise to readers of the *Times* (Figure 1.1).

Crookes argued that picric acid (aniline yellow), which had been used by synthetic dyers of silk and wool for over twenty years, was harmless even if the workmen who made it had skins "as yellow as guineas, and their hair of a beautiful green". The present problem, he suggested, might stem from the fact that manufacturers had recently taken to saturating picric acid with alkali before use. Consequently, if the wool was imperfectly washed, alkali would cause the irritation. If this were the explanation, he warned, manufacturers were in danger of blowing themselves up since alkaline picric acid was as explosive as nitroglycerine. One such factory had already been destroyed with the loss of life. In mock heroic fashion he said the sock wearers might vary the excitement of poisoning by exploding their toes instead! More seriously, he doubted that picric acid was really the culprit. Rather, it was probably the "Victoria Orange" and "Manchester Yellow" (dinitroparacresol, or 2,4-dinitro-1-naphthol discovered by Carl Martius in 1864) that had been rapidly developed commercially for rendering silk and wool a golden yellow, and the dinitroaniline, chloroxynaphthalic acid and nitrophenylenediamine used in making brilliant reds. Their "chromatic brilliancy" bore "no relation to the euphony of their names". He then offered to identify the poison if Webber and McVeigh would send him samples of the deadly socks.

Other *Times* readers poured out their troubles over coloured hosiery, leading "Barefoot of Taunton" to suggest that abandoning socks and stockings, as he had done, was the obvious remedy. Another English doctor practising in Le Havre then reported that he had warned about coloured socks the year before but his warning had been turned down for publication by *The Lancet*. The new feature of his precise observation was, that in the case he had seen, the weeping stripes on his patient's foot corresponded exactly to where the transverse stripes of red colouring were in the socks. The offending socks had been taken to the French government's laboratory at Rouen where they were professionally analysed. From analyst Bidard's report we learn that the coloured bands were in fact made of dyed silk that had been sewn into the violet ground of the woollen stocking. The violet of the main sock, which was the "aniline violet" first prepared by August Wilhelm Hofmann at the Royal College of Chemistry in London, was absolutely harmless, while the silk was dyed with fuchsine (aniline red). It was the latter that was causing the problem. Fuchsine had, in fact, been used in the dyeing trade for a good ten years, but hitherto had only been used in external clothing that did not particularly come into contact with the wearer's skin. In socks, however, the fuchsine was brought into close skin contact by shoe pressure. The French laboratory reported that because fuchsine was soluble in weak acids, it would therefore react with perspiration. The case of the poisoned socks seemed closed—it was due to fuchsine and sweaty feet.

The French intervention must have taken the wind out of Crookes's sails. He had offered to analyse the socks only to discover that it had already been done by a foreign government analyst. However, this did not deter Crookes, who completed the case a week later with a humorously written letter on poisonous dyes and a different conclusion. From the letters he had received, Crookes had been surprised how far the taste for gaudy hose had penetrated. There must be several hundred dozen pairs of these "chromatic torpedoes" in the public domain, he noted. Their male and female owners had no need to panic. There was really no reason why "young men and maidens should not continue to indulge in attire as startling and varied a colour as their good taste may permit". Rumours that the irritation was caused by arsenic poisoning were quite untrue. He had been assured by the largest aniline dyes maker in Europe (presumably the Atlas Works of Brooke, Simpson & Spiller in London's East End) that arsenic was no longer used in the manufacture of magenta. He identified the culprit as a new orange dye that was different from all the dyes he had previously dealt with. It was an acidic dye, insoluble in water, but soluble in alkalis. He had been unable to work out its composition and "sesquipedalian nomenclature".

From Crookes's brief description it sounds as if it was an azo-dye rather than fuchsine. On consulting Maurice Fox's wonderful compendium, *Dye-Makers of Great Britain*, it appears most likely that the offending dye was Field's Orange (also known as Field's Yellow), an amino-azo-benzene hydrochloride that Frederick Field prepared from aniline and developed at the Atlas Works in the 1860s. It is possible that the scare over poisonous socks in 1868 inhibited its use with textiles for, according to Fox, its main use was found in the coloration of varnishes and foodstuffs. Crookes put the risks in perspective. The number of cases of irritation had been few compared to the numbers of socks sold. The dye only affected wearers whose perspiration was alkaline (as opposed to the more normal acidic secretion to which the French had referred). Even so, he recommended that the particular orange dye's use should be restricted to articles of clothing that did not come into contact with the skin. There were plenty of harmless colourful dyestuffs, including phosphine (chrysaniline), aurine, Manchester Yellow, flavine and picric acid. Meanwhile, the hapless owners of poisonous socks should not throw them away. When washed with soap and soda they would "lose their stimulating action, both on the feet and on the optic nerve".

It was another twenty years before striped socks were made safe for sensitive skins when hosiery firms like Morley developed an oxidizing process that stabilised synthetic dyes. Widely advertised as "sanitary" or "hygienic dyes", hosiers were, at last, able to guarantee their coloured socks were stainless and proof against human perspiration.

Interestingly, a century later a closely related azo-dyestuff became implicated in another health scare. This was tartrazine, which had become widely used in the food industry as a colouring agent since its synthesis by Ziegler in 1884. Following the formation of the European Union in the 1950s, food additives that had passed stringent safety tests were identified by an E(urope) number. Tartrazine, for example, became identified on labels as E102. Synthetic colorants appealed to the food industry because they were cheaper than natural dyes, more stable, and usually more dramatic in their visual appeal. However, by the early 1960s there were a growing number of reports from parents who had noticed dramatic mood swings after their children had consumed brightly coloured sweets, cakes and soft drinks. A medical report published in 1975 identified colouring agents used in foodstuffs, as well as stabilising agents such as sodium benzoate, as a cause of hyperactivity and attention deficit in school children. It was not, however, until 2009 that the UK Food Standards Agency moved to ban such colorants from foodstuffs. Despite this recommendation, the European Food Safety Authority has remained adamant that E-number additives are safe. Nevertheless, most responsible food manufacturers have voluntarily removed substances like tartrazine from the food chain.

CHAPTER 2

Taste, Smell and Flavour

An essay written by William Prout (1785–1850) when he was a medical student at the University of Edinburgh shows that he was committed to a belief in the unity of matter several years before the publication of the two famous anonymous papers which contained "Prout's hypothesis" suggesting that all of the known elements might really be polymers of hydrogen. The student essay, *De facultate sentiendi*, a quarto manuscript of 26 pages is in English, despite the Latin title. It is an extraordinary example of the power and pitfall of analogical reasoning that is unguided by experiment. Prout's aim was to argue that sensation, like matter, was basically one thing. The five senses—touch, taste, smell, hearing and vision—were regarded by Prout as the peculiar sensations produced on specialized nervous apparatuses by the contact of a unique matter that was aggregated into the five different physical forms, solid, liquid, gaseous, ethereal and luciform. Tactile feelings, tastes, smells, sounds, and colours, he concluded, were all ultimately dependent "upon the different sizes, &c. of the aggregated particles of the same matter".

Prout's programme to reduce the number of sensations has never been realized, for in searching for analogies between sensations, he missed significant differences and over-emphasized superficial or accidental similarities. The notes on taste, smell and flavour, which closed his essay, are those that have best stood the

The Case of the Poisonous Socks: Tales from Chemistry
By William H. Brock
© William H. Brock, 2011
Published by the Royal Society of Chemistry, www.rsc.org

WILLIAM PROUT, F.R.S., F.R.C.P.
1785. —— 1850.

Figure 2.1   William Prout (1785–1850), a London physician who specialized in
the treatment of urinary and stomach diseases. Best known for his
speculation that the chemical elements were made from hydrogen,
he was also a pioneer of food chemistry. (RSC Library)

course of time. Here it is noteworthy that Prout's discussion was
based firmly on experience, and that these same notes formed the
basis for his first publication in a London medical journal in 1812
(Figure 2.1).

Prout had noticed that taste was commonly confused with
flavour, which physiologists believed to be the principle in a sapid
substance that excited taste. Instead, he suggested, taste and
flavour were physiologically different sensations. "Taste is that
modification of sensation which is caused by the contact of certain
substance soluble in water or saliva with the tongue, the nostrils
being at the same time closed and the tongue not being in contact
with any other part of the mouth". In his student essay Prout
mentioned for the first time experiments which supported his
opinions, in this case a definition of gustation. If substances like
nutmeg were placed upon the tongue and the nostrils closed or

plugged, then only a slight pungency was experienced. This was in fact the experimental justification for Prout's contention expressed in the main essay that the number of tastes was limited to acid, alkaline, bitter, sweet and "perhaps one or two more". In fact, nineteenth-century physiologists identified four tastes (sweet, sour, bitter and salty) so it is interesting (as we shall see below) that Prout should allow for the possibility of others. Since substances that excited taste usually also excited sensations when placed in contact with the body stripped of its cutis, he concluded incorrectly that all substances which excited taste were stimulants. But this fitted in with his conception that taste was similar to touch, even though it was the most limited and imperfect of all sensations.

What was usually taken for taste, said Prout, was flavour, which was really a combination of taste and smell. People with colds, who lost their sense of smell (*anosmia*) also lost their sense of "taste", *i.e.*, flavour; and people born without a tongue, or who had accidentally lost the tongue, could legitimately claim to "taste". "Smell", he wrote, "is that modification of compound sensation which is excited by various states of matter either in an aeriform state or in a state of extremely fine mechanical division when these are drawn in the air through the nose". The sensory mechanism of the nose, like the tongue, was chemical or galvanic. Flavour was another modification of sensation produced by the union of taste with smell. "Substances in general have the strongest flavour that are volatizable or partly soluble in air as well as water".

In his published paper Prout gave flavour the definition which has since passed into physiological literature. "Flavour is that sensation which is produced when substances under certain circumstances are introduced into the mouth, *the nostrils being at the same time open*". Olfaction was independent of gustation, but not of flavour. Olfaction could be influenced by strong flavours, as when vinegar was held in the mouth and masked the odour of ammonia held to the nose. The reverse experiment did not work, Prout explained, because ammonia had little flavour due to its great solubility in water. The seat of flavour was more extensive than that of taste and included the palate, the fauces and rear of the nose, the pharynx and the upper oesophagus.

As previously stated, these notes of taste, smell and flavour were published anonymously by Prout in 1812. Later the distinctions he had drawn were noted by his close friend John Elliotson (a fellow student at Edinburgh) and published in the latter's translation of Blumenbach's *Physiology* (1815). In this way Prout's analysis passed into the general literature of the physiology of sensation; curiously, however, his name has never been attached to it.

In the published account no hint of the original discussion on the nature of sensation was given; instead Prout added some remarks on the necessary conditions for the perception of the three chemical sensations. For taste, the tongue had to be moist and the sapid substance soluble in water or saliva. Metals, which were insoluble, excited taste because they were easily oxidized into soluble oxides. The sensation of taste was therefore chemical or electrical in character. "Taste appears to be universally produced by, or during the moment of, chemical action, especially when this is effected, as perhaps it is in every instance, by galvanic agency".

A third criterion for gustation was *motion*, for "if we apply any substance to the tongue in such a manner that it may be at rest, and its action circumscribed, the sense of taste will be very imperfectly, if at all produced". Once the substance was rolled about the tongue or the mouth, the taste became distinct. This condition continues to be demanded by modern physiologists since the motion of a substance over the tongue or around the mouth serves both the increase the rate of solution in saliva and to overcome the adaptation (*i.e.* the fatigue) of the chemical taste receptors.

Olfaction similarly required a moist nasal membrane, the solubility of the odorous substance in air, and the motion of the air through the nostrils. These conditions also remain accepted in modern physiology. Provided the air or gas which contained the odorous substance did not decompose it, its chemical nature was unimportant. If inspiration was momentarily suspended, the sensation of smell ceased.

Since flavour, according to Prout, was compounded of taste and smell, the same three conditions for its production were necessary. For this reason the most highly flavoured substances were those which were soluble in both water and air, "and hence all substances containing a volatile oil, as the various spiced, aromatics, &c. have this power in an eminent degree, though they excite comparatively little taste".

In the paper published in the *London Medical & Physical Journal* in 1812, Prout made only one significant alteration to his earlier account of flavour. Whereas in 1810 he had suggested that inspiration of air through the nostrils was not absolutely necessary to experience the sensation of flavour ("as flavour is equally strong during expiration and inspiration") in 1812 he criticized this viewpoint. Instead he thought that "during the act of *flavouring* … there is a slow *expiration* of the air through the nostrils, and at the same time generally a slow respiration through the mouth". Perhaps this afterthought came to him as a result of becoming expert in blowpipe analysis since this required the analyst to simultaneously expire air through the mouth and inspire air through the nostrils. Prout's point was presumable that substances volatilized in the mouth would find their way into the nasal cavities while moving in the direction of either the inspired or the expired air. Since olfaction occurs both when air is breathed in and out of the nose, flavour should be experienced under both conditions. Prout's unpublished opinion, then, seems to have been the better one. It has been said of olfaction that "owing to the lower path of the outward air, odour is usually noticed during inspiration, … [and] the many odours of food are mostly experienced through the vapours finding their way from the mouth and throat to the olfactory region during inspiration". The same may not be true of flavour since it is probable that the major part of our sense of flavour comes from the stimulation of the olfactory nerves by odorous air which is ejected when, during the act of swallowing food, the epiglottis closes the trachea. But whatever the complex origins of flavour may be, Prout's conditions of 1812 will not do.

The enumeration of the four tastes (sweet, sour, bitter and salty) was undoubtedly drawn from Europeans' eating and cooking habits. In 1907, however, the Japanese chemist Kikunae Ikeda (1864–1936) detected a quite different fifth taste in the dried seaweed soups to which he and other Japanese were partial. He described its savoury quality as *umani*, from a Japanese expression meaning "delicious". He decided that this unknown material was probably also responsible for the characteristic tastiness of meat, cheese, tomatoes and vegetables such as asparagus. Could he extract a common chemical from these different foodstuffs? (Figure 2.2).

Ikeda had graduated in chemistry in 1889 at the Imperial University of Tokyo, where the teaching was in English. In 1896,

Figure 2.2   Kikunae Ikeda (1864–1936), Japanese physical chemist who identified a fifth taste quality in 1908. He named it *umami* to separate it
from sweet, sour, bitter and salty. (Brock)

after teaching prospective science students at Tokyo's Normal
School, he was appointed an assistant professor at the Imperial
University. His head of department was the British-trained physical chemist, Jōji Sakurai (1856–1939) who was married to Ikeda's
sister. In 1899 Sakurai persuaded the Ministry of Education to
send Ikeda to Wilhelm Ostwald's laboratory in Leipzig in order to
gain first-hand experience of research in the new physical chemistry
pioneered by Ostwald, Arrhenius and van't Hoff. In Leipzig he
collaborated with Ostwald's student Georg Bredig (1868–1944) on
the catalytic action of colloidal platinum. In 1901, after a brief
sojourn in the Davy-Faraday Research Laboratory at the Royal
Institution in London, he returned to Tokyo as a full professor at
his alma mater.

Using his chemical expertise and a good deal of hard labour
evaporating vats of heated edible seaweed, he was able to extract
some brown crystals. A solution of these crystals tasted of *umani*
and proved to be the salt of an amino acid called glutamic acid,
$C_5H_8NHO_4 \cdot H_2O$ (α-aminoglutaric acid), a plant protein that
had been isolated by Heinrich Ratthausen in Germany in 1866.

Glutamic acid occurs in two optically active forms and Ikeda showed that it is only the *laevo* form of the sodium salt that gives the flavouring effect. Ikeda must have immediately realized the commercial value of his discovery since he patented the use of the monosodium salt of glutamic acid as a savoury essence in 1908 before publishing details of the extract in 1909. Ikeda's commitment to Ostwald's ionist form of physical chemistry undoubtedly helped him decide that monosodium glutamate, not glutamic acid, was the real source of *umani*.

Glutamines had previously been extracted from sugar beet by Ernst Schulze in 1883 and this proved a cheap and easy method to produce the condiment on an industrial scale. Together with a colleague, Saburo Suzicki, Ikeda set up the Ajinomoto Company (*Aji-No-Motu* is Japanese for "taste essence") in 1910 and monosodium glutamate (MSG) was born.

Although the condiment sold well enough in the Far East, it was largely ignored in the West. Partly this was due to the fact that Ikeda had published in Japanese; partly it was due to Western scepticism that there was a fifth taste; and partly it was due to Western indifference to Chinese and Japanese food. All this changed as a result of the Second World War, when American and European armies ate local food and experienced the enhanced flavour conferred by glutamate. Following the need for increased production in the 1950s, chemists developed a fermentation method to produce glutamate from molasses and wheat. However, it was not until 1995 that physiologists at the University of Miami showed that the human tongue had five different nerve receptors, the fifth being stimulated by glutamate molecules.

Ironically, French cooks may have discovered the value of the condiment at the time of Ikeda's discovery. The famous French chef, Auguste Escoffier, regularly enhanced the flavour of his recipes with veal stock for his gravy and sauces which, unbeknown to him, concentrated glutamates. His influential cookery book, *La Guide Culinaire*, was first published in 1903.

As Prout and Ikeda had foreseen, chemical knowledge and expertise would become essential in food production and research. By 1935, when the publisher George Newnes issued a fascinating four-volume work entitled *Chemistry in Commerce*, the food industry employed more chemists than any other branch of industry in Great Britain.

CHAPTER 3

Tales of Hofmann

August Wilhelm Hofmann (1818–1892) was the son of the government architect who enlarged Justus Liebig's laboratories at the University of Giessen in 1839 during the period when his son was studying law and languages at the same university. Attracted by Liebig's classes, young Hofmann abandoned jurisprudence for chemistry and obtained his doctorate with Liebig in 1841. For some years he remained at Giessen as Liebig's personal assistant, but in 1845, to further an academic career, he became a Privatdocent (lecturer) at the Prussian University of Bonn. In the same year, the Royal College of Chemistry was founded in London as a private school for the training of agricultural and industrial chemists. Following suitable guarantees of tenure in Germany if this British venture failed financially, Hofmann agreed to direct the English school, which was modelled on Giessen, for a period of two years. In the event, despite the College's financial insecurity until it was taken over by the British government in 1853, Hofmann stayed in London for twenty years, only returning to Germany, as professor of chemistry in Berlin, in 1865.

Hofmann's influence, as a teacher, on British and German chemistry, was striking. He continued the tradition of lucid teaching of analysis by laboratory instruction which had been established by Liebig, and created his own distinctive school of chemists who were interested primarily in experimental organic chemistry and its industrial applications. It is one of the amusing

The Case of the Poisonous Socks: Tales from Chemistry
By William H. Brock
© William H. Brock, 2011
Published by the Royal Society of Chemistry, www.rsc.org

Figure 3.1   August Wilhelm Hofmann (1818–1892), the Anglo-German pupil of Liebig's who directed the Royal College of Chemistry in London between 1845 and 1865 and who helped create the British dyestuffs industry. (Wellcome Images)

ironies of the history of chemistry that a chemist who was by all accounts ham-fisted when it came to chemical manipulation should have designed, developed and tested so much of the demonstration apparatus used in the teaching of elementary chemistry from the 1860s to the 1960s (Figure 3.1).

Hofmann was endowed with a sparkling intelligence and humour which made him a great favourite of London (and later) Berlin social gatherings. He was a good speaker, though in formal lectures or when writing letters, he often overindulged in literary embellishment and circumlocution. However, fame and fortune were won in the face of a private life that was continually clouded by domestic calamities—like the poet hero of Offenbach's *Tales of Hoffmann*. He was married four times and was three times a widower. Gossips said that he chose younger and more beautiful women each time.

To Hofmann in the 1860s it seemed obvious that a revolution was occurring in chemistry. Organic chemistry (the chemistry of carbon compounds) was being transformed from a dark forest or labyrinth into what Hofmann likened to a well-ordered city with streets, housing plots, market squares and factories. The purpose of his *Introduction to Modern Chemistry* (1865), the book which formed his leaving present to British chemists on the eve of his departure for Berlin, was to introduce order and rule into chemistry. This was to be achieved through the standardization of atomic and molecular weights and chemical formulae, and by the exploitation of a theory of types in which most chemical substances could be expressed as variations on the four prototype molecules—hydrogen, hydrogen chloride, water and ammonia. Once this was done, he argued, it became immediately apparent (as his friend and successor at the College, Edward Frankland, had also argued) that elements, or atoms, possessed fixed integral powers of combination or equivalence; a concept soon abbreviated to "valence" or "valency".

An excellent example of Hofmann's skills as a popular lecturer and of his ability to combine good chemistry with propaganda is the lecture on the dyestuffs mauve and magenta he gave at the Royal Institution in April 1862. In this account of the journey from coal to colour, which passed by any foul-smelling country (*i.e.* difficult chemistry), Hofmann used a variety of teaching aids including large wall-mounted diagrams, samples, models (including biscuit tins and wire cubes) and experimental demonstrations. His use of "structural models" (Hofmann's term) was to be perfected three years later when, in the same lecture theatre, he introduced coloured billiard ball models of atoms to help his audience understand the idea of valency. Such "glyptic formulae" as the Victorians called them soon proved of fundamental usefulness in teaching chemistry. By the 1870s they had become indispensable for understanding stereoisomerism. Atomic models were, perhaps, Hofmann's most important innovation in the popularization of chemistry.

The moral of this particular lecture on coal and colour was that mauve and magenta were Royal Institution colours insofar as benzene, their starting point, had been discovered there by Faraday in 1825. According to Hofmann, if anyone had asked Faraday at the time why he worked on benzene, he would have replied, "for the love of truth". Like many of his contemporaries, Hofmann paid lip service to the claim that pure research was the handmaiden of applied science. Ironically, in the same year, 1862, he was persuaded otherwise by his experiences as a juror for the 1862 International Exhibition in London. Through the stimulus of the work on synthetic dyestuffs of his pupil, Edward Nicholson, Hofmann became an advocate of a symbiosis between pure science, chemical industry and trade. Pure academic research, he concluded, would in the future be as much indebted to the work of industrial chemists, as it would to university science. It was this message that he carried back to Berlin in 1865.

As Heinrich Caro said in concluding his 1892 history of the dyestuffs industry, Hofmann had been the loyal gardener who, for fifty years, had tended the great tree of academic and industrial chemistry.

CHAPTER 4

Liebig on Toast

In 1821, fresh from his sensational exposal of the horrifying extent of contemporary food adulteration, and his even more sensational dismissal from the Royal Institution's employment for defacing library books, Frederick Accum staved off starvation by compiling a pot-boiler entitled *Culinary Cookery*. "The subject may appear frivolous", Accum admitted, "but let it be remembered that it is by the application of the principles of philosophy to the ordinary affairs of life, that science diffuses her benefits and perfects her claim to the gratitude of mankind".

The most significant bearer of this message in the mid-nineteenth century was the famous German organic chemist, Justus Liebig (1803–1873). At the University of Giessen in the 1830s, Liebig perfected the methods of organic analysis and set up a model teaching and research laboratory that attracted students from all over the world. When he moved to Munich in 1852 he ceased teaching, preferring instead the role of science popularizer and expert (Figure 4.1).

In a reminiscence, Liebig tells us that the kitchen was where his career had begun—where he learned French in the royal kitchen from the wife of one of the Duke of Hesse-Darmstadt's cooks. Fascinated by the food preparation that he witnessed, he retained a taste for cooking, and in leisure hours occupied himself with "the mysteries of the kitchen". In fact, he gave more than leisure hours in this pursuit as he followed the success of his book, *Animal Chemistry* (1842), with an elaborate study of meat in *The Chemistry*

The Case of the Poisonous Socks: Tales from Chemistry
By William H. Brock
© William H. Brock, 2011
Published by the Royal Society of Chemistry, www.rsc.org

Figure 4.1   Justus von Liebig (1803–1873), German chemist who created the extract of meat industry in South America. (Wellcome Images)

of Food (1847) and in the third English edition (1851) of his popular *Chemical Letters*.

Liebig argued that the essential nutrients in meat were not contained in the muscle fibres of the flesh, but in the muscle fluids which were easily lost in boiling and roasting. It was essential, therefore, to minimize the loss of these vital salts if nutritional quality was to be maintained. The way to do this (apart from always using stock as gravy or soup) was to sear the surface of a joint by plunging it into boiling water, so that the albumin (protein) was coagulated sufficiently to prevent leakage of the internal juices. Contemporary cookery books of the 1850s, such as Eliza Acton's *Modern Cookery*, quickly adopted this misleading but "scientific" advice and, one suspects, applied it to cooking vegetables as well!

One of Liebig's closest English friends was the Liverpool–Irish alkali manufacturer, James Muspratt (1793–1866). In 1854, Liebig and his wife were hosts to Muspratt's seventeen-year-old daughter, Emma, who had come to Munich to learn German. Unfortunately, Emma caught scarlet fever and lay at death's door unable to eat until Liebig miraculously restored her to health with an "extractum carnis", *i.e.* meat extract.

Following this incident, Liebig encouraged several German and British pharmacies to prepare the remedy for use in hospitals and by invalids. But production costs were high since pharmacists had to buy meat directly from local slaughter houses. The answer, as Liebig recognized, was to exploit the wasted carcasses of the hordes of cattle that were reared in Australia and South America for their hides and for tallow production.

Although Liebig openly negotiated with several cattle barons in Australia, it was not until the 1860s that a meat extract industry emerged. In 1862, fresh from building railways in Uruguay, the German engineer, Georg Giebert, propositioned Liebig to invest in the purchase of a *salderos* (beef salting plant) at Villa Independencia (later Fray Bentos) on the River Uruguay, and to convert it into an extract-processing and canning plant. The Liebig Extract of Meat Company was launched on the London Stock Exchange in 1865.

With meat refrigeration still twenty years away, the company (and its rivals) hit a jackpot: for it was the decade of rising European population, in which cattle plague was rife, and subsequent sanitary investigations revealed that the poor were able to afford less than a pound in weight of meat a week. The extract could, therefore, be advertised and sold not just as a medicine, but as a cheap nutritious food (Figure 4.2).

By the 1870s, however, following a better understanding of physiology and energy, meat extracts were shown to have little nutritional value. It was, at best, only a stimulating condiment. The Liebig Company was therefore forced to change its advertising tactics. Accordingly, Liebig asked the famous German cookery writer, Henriette Davidis, to devise a series of recipes using the extract. Her *Kraftküche von Liebigs Fleischextract* proved so successful that the company commissioned cookery books in other languages, including Hannah Young's exquisitely illustrated *Practical Cookery Book* (1895). A woman with outstanding

Figure 4.2   Advertisement for Liebig's beef wine, another commercial version of his extract of meat used as a medicinal tonic and restorative. (Wellcome Images)

entrepreneurial abilities, Young outstripped the fame of her husband. To retain proprieties, it was he who changed his name to Young on their marriage. Every savoury dish in her book uses some of Liebig's extract, including the fish recipes. Only the sweets escaped attention. Included were recipes for "Liebig sandwiches" and, of course, "Liebig on toast".

Within a decade after Liebig's death in 1873, Liebig's Company had become one of the largest cattle-farming enterprises in the world and renowned for two other beef products: Fray Bentos Corned Beef, which was launched in 1879, and Oxo. The latter, first marketed in 1900, reached its familiar solid cubic form in 1911 when it was cleverly advertised as "the penny a cube". Liebig's work on food also led him to improve the making of bread and coffee, and to develop a baking powder and babies' milk. It encouraged other entrepreneurs to develop dried soups and sauces, and to see the chemist not merely as an analyst and quality controller, but as someone who could bring chemistry from the kitchen into the food laboratory (Figure 4.3).

Figure 4.3   View from an office at Waterloo station in the early 1900s capturing advertisements for 'Oxo', a new solid, cubic form of extract of meat made by the Liebig Company, and for 'Bovril', a meat extract developed by a rival company. (Science & Society Picture Library)

CHAPTER 5

The Future of Research at the Royal Institution (London) and the Smithsonian Institution (Washington)

From time to time public and private institutions are faced with a moral dilemma over whether or not to accept or reject a financial endowment or gift in kind, either because it comes with strings attached or because it may conflict with one of the institution's principles. For example, in the 1850s University College London was embarrassed by the Peene Bequest. In his will, dated 9 March 1853, Dr William Gurden Peene, a Kent doctor, left £1,700 to the College for the purchase of library books. The snag was that the books had to be chosen by the three professors of Greek, Latin and Mathematics, providing they were members of the Church of England. If they were not (and Augustus De Morgan, the Professor of Mathematics was not), the College Council was permitted to appoint any other professor or College alumnus who was. Following embarrassing discussions over whether acceptance of the gift would infringe the College's principles of religious equality, Council decided to accept but its decision nearly prompted the resignation of De Morgan from his College chair.

The Case of the Poisonous Socks: Tales from Chemistry
By William H. Brock
© William H. Brock, 2011
Published by the Royal Society of Chemistry, www.rsc.org

No doubt many institutions have similar stories to tell. In 1831 a casual visitor presented the Society of Arts with a sealed copy of the will of a medical practitioner, Dr George Swiney. The Society was unable to discover Swiney's address or identity him until 1844 when it learned that Swiney had died. He had left the Society £5,000 on condition "that a sum of £100 contained in a silver cup of the same value should be awarded every fifth anniversary of the testator's death as a prize for the author of the best published book on Jurisprudence". Not only had the Society of Arts no interest in law, but it found itself bemused by a further condition that the adjudicators of the prize were required to be members of the Society *and* of the Royal College of Physicians, together with "the wives of such of them as happen to be married". In this instance honour was satisfied by making the award for Medical and General Jurisprudence in each half decade, with the Royal College of Physicians and the Society of Arts adjudicating alternatively.

FINANCING THE ROYAL INSTITUTION

The Royal Institution was founded in 1799 without endowments. Its finances were modelled on the tried and tested methods used for establishing and running voluntary hospitals, and charity and Sunday schools. In exchange for an investment of fifty guineas or more, the founding proprietor members could obtain tickets for themselves, their friends and their servants to attend the Institution's activities and to use its facilities. Although this proprietorial system was abandoned at the suggestion of Humphry Davy as early as 1810, when annual subscriptions were introduced, the Royal Institution continued without endowment for two more decades. Subscription income, as well as some of the privately earned and government-sponsored research monies earned by William Brande and Michael Faraday had to cover fabric expenses, professorial and assistants' salaries, as well as laboratory expenses.

Until 1863, when the Royal Institution's Secretary, Henry Bence Jones, launched a Fund for the Promotion of Experimental Research, its only endowment had come from the eccentric John Fuller in 1833. This endowment established the Fullerian Chairs of Chemistry and Physiology with honoraria of £95 per annum.

Interest from a legacy of £1,000 left by Mrs Hannah Acton in 1838 was not available for research purposes since it had to pay for a septennial essay prize on natural theology. When James Dewar, who had revived Jones's Fund in 1886 for purposes of low temperature research, gave his Presidential Address to the British Association for the Advancement of Science at Belfast in 1902, he made a plea for greater public expenditure on science. He publicly audited what it had cost the Royal Institution to be what an American writer had described as an institution that had done more for British science in a century "than all the English (*sic*) Universities put together". In a century, expenditure had been:

Professors' and assistants' salaries = £76,190

Laboratory expenses = £24,430

Total = £100,620

If to this figure one recognized the nation's benevolence in granting Faraday a Civil List Pension between 1835 and 1867 (£9,600) and the £9,580 subscribed through Bence Jones's Experimental Research Fund, Dewar calculated that for a mere total public expenditure of £120,000, or roughly £1,200 per annum, a century of British science had been gloriously advanced.

Of course, as he spoke, during the previous six years the research endowments of the Royal Institution had been transformed by Ludwig Mond's covenant of the annual interest from a capital long term endowment of £62,000 in 1894. Originally destined to be made over to the Royal Institution in 1926, this capital sum was, in practice, transferred to the Institution as War Bonds in 1918, by which time the capital sum was £66,500.

Less well known and indeed completely forgotten, since the gift has long since ceased to be commemorated within the Royal Institution's accounts or lecture programmes, is the Trust Fund given to the Royal Institution in 1892 by the Anglo-American businessman, Thomas George Hodgkins (1803–1892). Yet it was the annual interest from this endowment, coupled with the revived Bence Jones Fund, that enabled Dewar in the 1890s to engage in expensive low temperature physics—what the chemist Henry Armstrong called the exploration of the hyper-arctic.

ANOTHER ECCENTRIC DONOR?

Who was Thomas George Hodgkins? He had been born in England in 1803 and gone to sea with the East India Company at the age of seventeen when his father remarried and Hodgkins decided that he could not get on with his stepmother. Following a shipwreck and near loss of his life, he found himself in a Calcutta hospital. We are told that he resolved there and then that if his life was spared he would devote himself to the welfare of mankind. He returned to Britain, married and emigrated to New York in about 1830 where he made a fortune manufacturing confectionery. A millionaire by 1859, he apparently travelled extensively in Europe on at least four occasions. A surviving letter of July 1866, written when he was revisiting London, expressed delight at metropolitan improvements, the absence of drunkenness and the fact that working class life had been improved through parks and free museums.

On his retirement in 1873 he bought property at Setauket on Long Island in New York State which he named Bramblelye Farm. In his late-eighties, aware that his end was near, he began to dispose of his entire estate of half a million dollars in generously large amounts: £20,000 to the New York and Brooklyn Societies for Prevention of Cruelty to Children; £20,000 to the American Society for the Prevention of Cruelty to Animals; a Public Library for his local community in Setauket; and £6,000 to his doctor, M. L. Chambers, to look after him for the remainder of his life. Chambers was to lease Bramblelye Farm from 1890 (Figure 5.1).

So far, these legacies will sound like the conventional actions of rich American and British Victorian philanthropists. However, in November 1891, Hodgkins donated $200,000 (£40,000) to the Smithsonian Institution in Washington with the seemingly odd proviso that, although half of the money might be applied to the general purposes of the Institution, the remaining $100,000 was to be used for "the increase and diffusion of more exact knowledge in regard to the nature and properties of atmospheric air in connection with the welfare of man". It took Samuel P. Langley (1834–1906), the Smithsonian Institution's third Secretary, and the Trustees of the Institution, two years to work out how to diffuse this largesse in ways that would genuinely forward scientific research in line with the donor's intentions. The outcome was a series of international prize essay competitions that were to be

Figure 5.1   Thomas George Hodgkins (1803–1892), American confectioner
and philanthropist who endowed money for research at the Royal
Institution in London and the Smithsonian Institution in
Washington. From a posthumous portrait made by Robert George
Hardie in 1893 now in the National Portrait Gallery in Washington.
(Smithsonian Institution Archives)

rewarded by financial prizes, and a series of gold, silver and bronze
Hodgkins Medals. Even more complicated legally was the fact that
Hodgkins made the Smithsonian Institution his residual legatee.
These further endowments (some $50,000) were placed in the
general Hodgkins Fund when they eventually became available
in 1907.

THE ROYAL INSTITUTION REACTS

By then Hodgkins's name was well-known to Dewar and the Royal
Institution's Managers. This was because the final $100,000 of
Hodgkins's fortune was offered to the Royal Institution as a Trust
Fund in September 1892, providing certain conditions were met. In
a letter to Sir James Crichton Browne, the Royal Institution's Vice-
President, Hodgkins wrote:

> "It is generally admitted by the two branches of Anglo-Saxons
> [Britons and Americans] that we have erred, strayed like lost

sheep, of which no doubt can possibly be entertained; it may be worth considering how often this admission, frankly conceded, will enable us to find our way back to the fold?"

As a former director of a public lunatic asylum and an active Commissioner of Lunacy, Crichton Browne possibly recognized symptoms of dementia, obsession and ennui in these words. But we must read on:

> "Although I have no claim to be considered as a scientist, I reverence science; and I recognize that the sum totals of all science be embraced in an accurate knowledge of the mutual relations that exist between man and his creator. Outside of these two factors, the idea of science, cannot be entertained; it is what Herbert Spencer would characterize as "unthinkable". My Funds are invested in a Trust Co., payable to my order on demand; but before I surrender them, I must be satisfied that they be used in the *true* cause of science; coupled with the feeble effort to retrieve the threatened disaster at Dorking, destined I fear to prove the Thermopyle of the noble Norman element."

To interpret this seemingly demented statement, we must remember that General George Tomkyns Chesney (1830–1895) had published the sensational essay, 'The Battle of Dorking' in *Blackwoods Magazine* in 1871. Sometimes described as the most remarkable short story published in the nineteenth century, Chesney's military account of an imaginary invasion of Britain by the French had exposed Britain's lack of military preparedness. The essay had proved a best-seller in America as well as in Britain.

A Deed Poll accompanying Hodgkins's letter to the Royal Institution undertook to apply funds according to the Trust Deed providing that the income was applied:

> "... exclusively in investigation of the relations and co-relations existing between man and his Creator. It is the belief of [Mr Hodgkins] that with the present generation God has almost been forgotten and he appeals to the scientific men of the present and coming centuries to whom we look to direct thought to lead us back to the source of all knowledge."

The gift (a James Smithson donation in reverse) was, therefore, reminiscent of the donation by Mrs Acton of fifty years before, but worth twenty times in amount. There was no time to negotiate. Hodgkins's secretary and legal adviser told Sir Frederick Bramwell, the engineer Secretary of the Royal Institution, that Hodgkins's life hung by a thread. Decide immediately whether to accept, Bramwell was told, or forego the offer. Bramwell accepted with alacrity by telegram and humbly informed the President of the Royal Institution, the Duke of Northumberland, that he trusted he had done the right thing in not consulting the Institution's managers. The Duke agreed that Bramwell had done the right thing; after all, even though Hodgkins's conditions seemed embarrassingly religious, they permitted "the widest and most general interpretation".

THE SMITHSONIAN INSTITUTION REACTS

Although the Royal Institution never inquired of the Smithsonian Institution, it could easily have been reassured by its American sister's experience. For the Smithsonian negotiations, beginning in 1891 a year before those of the Royal Institution, had been conducted at a more leisurely pace. Both the Assistant Secretary, George Brown Goode, and the Secretary, Samuel P. Langley, had personally visited Hodgkins, as well as carrying out a correspondence with him. While appreciating that he was a little eccentric, they concluded that he was not senile and that his views had been nurtured by wide reading and philosophical rumination. Convinced of the elderly man's sincerity and sanity, the Smithsonian trustees duly reported that while half of Hodgkins's donation of $200,000 could be used for general research and maintenance purposes, the income from the other $100,000 was to be deployed as soon as possible for "the increase and diffusion of more exact knowledge in regard to the nature and properties of air in connection with the welfare of man". In this connection, prizes were to be awarded for three different essay competitions, the closing date for which was to be July 1894:

1. $10,000 for an essay announcing a new discovery concerning air
2. $2,000 for the best essay which applied the known properties of air to any other area of scientific investigation
3. $1,000 for a popular treatise on air.

In addition, the Trustees announced the annual or biannual award of a Hodgkins Medal for a scientific treatise on air.

A series of undated notes and press cuttings in the Smithsonian Institution's Hodgkins's file might suggest senility, but actually reveal typical fin-de siècle thought processes.

"Without man, as a primary factor, what becomes of Terrestrial science, or religion or conscious intelligence? Do a few pounds of earthy salts and a few parts of water constitute man:- or the home he lives in? Is he an ethereal being? insolubly connected with his Creator? He blew into his [man's] nostrils the breath of life, and man became a "living soul". Does "living soul" mean nothing? Or does it mean something:- and if it really meant nothing: what the "doose" [*sic*] could the prophet have meant? But suppose the soul endowed with life, it must have needed sustenance and atmospheric air was just the pabulum required ..."

Another scrap suggests that his mother had been consumptive, thus leading him to paranoid reflections that American civilization was threatening its future by sealing up apartments, removing natural ventilation and air in homes, theatres, schools and street cars. Carbon dioxide was slowly poisoning man.

In drafts of letters to the editor of New York's *Evening Post*, written (as he says) at the age of 76, he berated Manhattan for being disease-ridden and full of lunacy instead of the healthiest city on earth. The fault was educational:

"What should be the highest ambition of a citizen, or of a State—is it to raise men that are men: or should they as a matter of greater convenience and economy import them already made [*i.e.* by immigration]? To this latter policy there is one slight objection: if we raise a population deficient in the *vis viate* necessary to do honor to the Republic, we run the danger of creating one that has no other vocation than that of preying upon it. Therefore, if we are to place our chief reliance on the imported article, may it not [be] worth while to consider the policy of legalising infanticide, to rid ourselves of the inconvenience of rearing a population of those unfortunate beings who cannot work. Children may be made in the dark;

but to turn out *men* equal to the standard of those who
founded this republic, and who fought to establish it, it might
be advisable to hunt up the original recipe in the book of
Genesis 'and when found make a note of it'."

Such racial views concerning the degeneracy of society were by
no means uncommon in America or Britain in the 1880s. For
Hodgkins, the solution lay in science and in an understanding of
the true relations between science and religion. As he told William
Jay Youmans in 1887:

> "I reverence (practical) science, and its enlightened Professors.
> Whereas Ministers of religion ought to teach the laws of God
> in Exodus XX [*i.e.* the ten Commandments], scientists ought
> to enforce by argument and persuasion such laws of God as
> were omitted in the record [*i.e.* the Bible]. Science and religion
> have to come together again. They must not be considered
> antagonistical; synonymous or interchangeable; but as corre-
> latives as the right arm and the left; or the knife and fork; to be
> used conjointly."

When these quoted notes and scraps were obtained by the
Smithsonian as part of Hodgkins's legacy in May 1894, Dr M. L.
Chambers (Hodgkins's physician) observed dryly that though they
might reveal Hodgkins's personality they could be "misleading to
anyone who had not enjoyed his acquaintance".

Langley would have agreed, having enjoyed the old man's hospitality and company. As Hodgkins told Langley in June 1891:

> "I cannot resist the temptation of putting on record, that with
> Thomas Carlyle, my belief is in God as an entity, not as a
> symbol. I believe that my donation, to the Smithsonian
> Institution, is a surrender of a trust held from the Almighty."

Early that month Goode, the Assistant Secretary, had reported to
Langley that he had found Hodgkins:

> "... to be a dignified, genial, courteous, and altogether
> delightful old gentleman. He is feeble, and suffering from some
> of the infirmities incident to old age, but bright, quick of

comprehension, and unusually clear and vigorous in mind for a man who is nearly ninety ... He thinks mankind is degenerating and that this due to lack of regard for pure air. He is full of anxiety to do something for the welfare of the human race, and is deeply in sympathy with the objects of the Smithsonian Institution. Indeed he said that he might well have been its founder if he had been born in time ... He looks upon pure air with feelings akin to worship—through it, he hopes that the man of the future, who uses it properly, will perhaps be able to in some way communicate directly with his creator. He told me it was 'the breath of God'."

Like Hodgkins's doctor, Goode concluded that Hodgkins was sane. On the other hand, and no doubt recalling the Tichborne claimant, Chambers obviously disapproved of any part of Hodgkins's fortune returning to Britain.

"It might be the means of working up a host of English heirs who would trouble his estate for a long time and probably succeed in carrying that to England also."

WHAT THE ROYAL INSTITUTION DID WITH THE MONEY

The Royal Institution's money arrived in mid-October 1892 and exchanged as the sum of £20,523 sterling, which was immediately invested in $2\frac{3}{4}\%$ Consols. By happy coincidence the Goldsmiths Company donated £1,000 for low temperature research the same month. Things were looking brighter for James Dewar following the Royal Society's refusal to support his work from its research fund—a refusal for which Dewar never forgave the Society's Secretary, the physiologist Michael Foster (Figure 5.2).

The interest from Hodgkins's gift was to remain his during his remaining lifetime, but he promptly died on 25 November 1892. Although there were further legal complications with the Inland Revenue (which claimed £2,862 in death duties despite Hodgkins's American citizenship), and with Hodgkins's estate (which demanded a month's interest of £65), by March 1893 Dewar had a research income of over £500 per annum from the Hodgkins Fund

Figure 5.2   Sir James Dewar (1842–1923), Scots chemist who directed the
laboratories at the Royal Institution between 1877 and 1923. His
expensive low temperature research on the liquefaction and solidification of gases was financially supported by Hodgkins. (RSC Library)

to add to what now became known as the Low Temperature
Research Fund.

However, the managers still faced a moral problem that had
already been addressed by the trustees of the Smithsonian: how
could the Institution justify directing the money towards low
temperature physics? By good fortune, of course, Dewar was
working on atmospheric air—albeit he was attempting to liquefy it.
In December 1892, on Dewar's advice, the Royal Institution
managers resolved that the Fund could legitimately be applied to
the attainment of absolute zero, *i.e.* to "the attainment of truth",
which is what science was all about, and what Hodgkins had
described as "the investigation of the relations and co-relations
between man and his Creator". Further, using the established
model of the Acton Bequest, they agreed that every seven years
they would award someone with a hundred guineas to review the

development of the low temperature work to which the fund was being applied. The Smithsonian's solution to the Hodgkins's bequest was to be rather more elaborate, but before turning to this, let us take the Royal Institution story a little further.

The first reviewer of the Royal Institution's use of the Hodgkins's bequest became due in 1901, when the astronomical writer, Agnes M. Clerke, who had received the Actonian Prize in 1893, agreed to undertake the review. Her twenty-page essay on low temperature physics at the Royal Institution between 1893 and 1900 duly appeared in the *Proceedings of the Royal Institution* in 1901. It was attractively illustrated, but apart from some perfunctory references to the laws of nature as formulating intelligent purpose, it was a straightforward account of Dewar's liquefaction and solidification of hydrogen and associated researches. Clerke's essay was also very discreet and nowhere referred to Dewar's feuds with William Ramsay at University College or his contretemps with the inventor William Hampson over the apparatus for the liquefaction of air.

Seven years later, Henry Armstrong, a sometime manager of the Royal Institution, member of the Davy-Faraday Laboratory Committee, and close personal friend of Dewar's, was asked to write the review for the years 1900–1907. He did so slowly and it was not published until 1910. Like all Armstrong's work, however, it was quirky and richly entertaining.

Armstrong had been associated with the Royal Institution since 1867 when, as Frankland's research assistant, he used gas analysis equipment in its basement laboratories. From here he graduated upstairs to hear all the great public lecturers of the day such as John Tyndall. Both Ramsay and Armstrong had been contenders with Dewar for the Fullerian Professorship of Chemistry in 1877, but whereas Ramsay became Dewar's mortal enemy and rival, Armstrong was one of the few contemporaries with whom Dewar got on well. Armstrong described Dewar as "choleric, irascible and a good hater", and of how fascinating he found him. Indeed, he was so fascinated by Dewar's character that he tried to study him as a psychological specimen, but was never able to break through the external shell to reach the human being below. Even so, they hit it off; they shared obsessions about the quality of scientific education, the decline in scientific ethics (as they saw it), and they both had a mordant suspicion of William Ramsay, Michael Foster and their Royal Society cronies. Armstrong was therefore a very good choice as far as Dewar was concerned to write the septennial review of his laboratory work and to paint Dewar's output as whiter than snow.

Like Agnes Clerke, Armstrong ignored reality, although the 1907 essay on what he called the Royal Institution's "Charcoal Period" of low temperature research, did contain several digs at Ramsay. In a remarkable section on the "Future of Research at the Royal Institution" he also noted that the Institution was still the only venue in the United Kingdom of Great Britain and Ireland where public recognition was given to the value of science. Unfortunately, Ludwig Mond's expectations in providing facilities in the associated Davy-Faraday Research Laboratories had not been fully realized because Britain still neglected science teaching in her schools. An obsession with examinations had also all but destroyed the originality of those amateurs who might wish to take bench space in the Davy-Faraday Research Laboratories. These well-known hobby horses of Armstrong allowed him to make a plea for the appointment of more research assistants at the Royal Institution and greater public help for its work.

The First World War began a new era in the development of the Royal Institution and with it the end of the Hodgkins Trust. In July 1914, when they invited Armstrong to make the third septennial review of its expenditure, the Managers decided that it should be the last: "The revenue derived from the Hodgkins Trust is not now being allocated to meet specific expenditure in Chemical and Physical Research Problems; but is being used to aid the efficient working of the Institution in all its branches". The move was the only possible one given that Dewar had abandoned the hyper-Arctic for the delicate mysterious world of soap films. With the appearance of Armstrong's final essay in the *Proceedings* for 1916, Thomas Hodgkins's name disappeared from the records of the Royal Institution. It is appropriate to remember him again in the British context and to record that his legacy continues to "lead us back to the source of all knowledge".

THE AMERICAN FUND

Hodgkins's name has, however, been perpetuated at the Smithsonian Institution. It is interesting to recall that William Ramsay persuaded his colleague Lord Rayleigh to hold back on the public announcement of the discovery of argon in 1894 in order to enter the first international Hodgkins Prize essay competition for $10,000, which they duly won in August 1895 after 218 essays had been read. This award would undoubtedly have pleased Hodgkins personally, for Ramsay and Rayleigh had said something decidedly original and hitherto unknown about the nature of atmospheric air. Moreover, it was a discovery that was to have a dramatic effect upon the ordering of elements and upon ideas concerning the composition of matter. In turn, James Dewar (Ramsay's rival and bête noire) was to be awarded the first of the Smithsonian Hodgkins Gold Medals in 1899 for "his exceptionally important contributions to our knowledge of the nature and properties of atmospheric air, and the practical application of them to the welfare of mankind". Another $1,000 was awarded to Henri de Varigny of Paris for the best popular treatise on air.

The second prize of $2,000 was not awarded from lack of entries. However, a number of the contestants who received honourable mention, like the English chemist and spectroscopist E. C. C. Baly, were encouraged to apply for small research grants. Baly received £150 in October 1896 to further his investigation of the expansion of oxygen. And over the next decade or so, the Hodgkins Fund dispersed grants internationally for work ranging from cloud condensation, photographs of lightning, the flight of birds and insects, architectural acoustics, a speech machine, the rare gases (M. W. Travers), tuberculosis and aeronautics. While continuing initially to support essay research projects that had something to do with air (this, conveniently for Langley, could include aeronautics), the Institution issued only two Hodgkins Gold Medals during Langley's lifetime—those to Dewar in 1899 and to J. J. Thomson in 1902 for his work on the conductivity of gases. No further medals were issued until 1965 when the original medal was redesigned and redesignated as an annual or biannual award for environmental research. A Hodgkins Silver Medal was also struck in 1894, but never awarded. Instead it was given to Pembroke College, Oxford, in memory of James Smithson, the former graduate of the College whose unexpected endowment to the American people in 1829 had created the Smithsonian Institution.

CONCLUSIONS

The story told above may seem a minor one in the history of science and its institutionalization, but the Hodgkins's donation was undoubtedly of major significance to the two institutions concerned at the time of the respective endowments. What is interesting is in how the two institutions on different sides of the Atlantic dealt with an offer of money that outclassed anything that either had ever been offered before. Both institutions suspected that the donor was a little mad. The Americans were, however, in a position to investigate Hodgkins's state of mental health and decide that his offer was that of a logical mind. The English on the other hand, who were desperate for money to push forward an important programme of research on the liquefaction of air, took the money and asked no questions. The Americans may well have been suspicious, and not a little jealous, that some of Hodgkins's fortune was going to the British, but may have thought better of advising Hodgkins against this gift in view of Hodgkins's original nationality, as well as the memory of James Smithson's generous bequest to America in the face of protests from the British.

It could be argued that the British actually used the Hodgkins Fund more effectively than the Americans at first did. Although $100,000 was ploughed into the general Smithsonian capital for the support of science, Langley and his colleagues were in fact forced to think of ways of using the interest from another $100,000 immediately in support of Hodgkins own private obsession with the quality of air. There was, perhaps, little option but to go down the path of monetary prizes for work done, and the occasional retrospective medal for outstanding research in the phenomenon of atmospheric air broadly interpreted. Once the initial competition had been held however, the Americans were able to use the Hodgkins Fund quite effectively in the award of small grants for research projects connected with the atmosphere or with the properties of air and its constituents. Langley, his successors and their advisors proved effective in their choices, and although it cannot be said with conviction that any major discovery resulted from such funding, much was learned concerning atmospheric physics, aeronautics, and the chemistry and physics of air from such sponsorship. By the same token, Dewar's research on the liquefaction of hydrogen, and the failed attempt at the liquefaction of helium, were enabled by the extra funding his low temperature researches received from Hodgkins.

By the time of the First World War, however, on both sides of the Atlantic, the decision was taken quietly to allocate the Hodgkins Funds to the general purposes of their respective institutions. In the case of the Royal Institution, the name of Hodgkins disappeared from the balance sheets completely; at the Smithsonian Institution, while the Hodgkins Fund continued to appear in the balance sheet, the fund was no longer specifically directed towards research on the atmosphere but ploughed directly into astrophysical research. The award of one or more Hodgkins Medals seems to have lain forgotten until 1965 when the decision was taken to revive the award. The Trustees then sensibly decided to make the award recognition of "important contributions to knowledge of the physical environment bearing upon the welfare of man". Annual or bi-annual awards were accordingly made from 1965 to 1986. By then Hodgkins's capital was still producing an annual income for the Institution of $6,960. Since then, for whatever reason, awards of medals have again become intermittent.

One final delicious irony remains to be noted. It may well be that Hodgkins had never intended that part of his fortune should go to the Royal Institution. One valid reading of Hodgkins's, his doctor's and his lawyer's surviving correspondence is that he had intended the money to go to the Royal Society, but that due to his lawyer's unfamiliarity with British scientific institutions, and plagued by the resonance of the Smithsonian Institution's title, he telegraphed the Royal Institution by mistake. What the Royal Society might have done with Hodgkins's money must remain a conjecture.

CHAPTER 6

The Future of Chemistry in 1901

There are many justifications for being interested in, and investigating the history of chemistry. At the end of the day, however, they boil down to humanizing science teaching and research. The great German historian of chemistry, Hermann Kopp, put this well in the 1880s when writing about alchemy. While he dismissed the alchemists' elixir of life as an idle fancy, he saw the history of chemistry as the next best alternative. Through history we live and experience other lives vicariously and thereby extend and enrich our own lives.

As it happens, Kopp died at the age of 75 in 1892 having lived himself through a truly remarkable century of chemical change. A British pupil, Edward Thorpe, gave the Chemical Society's Memorial Lecture in February 1893 (curiously the Society only memorialized its most distinguished foreign members and never its own countrymen). Thorpe made it clear that the future of a great deal of chemistry lay in pursuing Kopp's search for meaningful mathematical or colligative, additive and constitutive laws of the physical properties of the elements and their compounds. Today we may not think Kopp's research programme as particularly important and even rather dull. But there can be little doubt that chemists 150 years ago viewed it as a leading research activity and one that would help unveil the ultimate properties of matter. When, for example, Sydney Young was President of Section B (the Chemical Section) of the British Association meeting in Southport in 1904, he devoted his

The Case of the Poisonous Socks: Tales from Chemistry
By William H. Brock
© William H. Brock, 2011
Published by the Royal Society of Chemistry, www.rsc.org

entire address to a review of Kopp's work and his own extensions of it measuring the vapour pressures, specific volumes and critical constants of pure liquids. All this was done with a view of testing van der Waals's generalizations regarding corresponding temperatures, pressures and volumes. Later he went on to detect relationships between physical properties and molecular structures in homologous series and with different functional groups.

To chemists working around 1901, then, it must have seemed that the future lay in physical chemistry. (It was about this time that Fritz Haber, himself trained as an *Organiker* before turning to physical chemistry, put the natural products chemist Richard Willstätter down with the remark that organic chemistry was finished unless it introduced physical methods and theory.) Indeed, it was at the turn of the twentieth century that Walther Nernst showed the significance of thermodynamics in his derivation of an expression for the free energy of reactions which Fritz Haber was to use in the investigation of gas equilibria. Haber's investigation of the fixation of atmospheric nitrogen by catalytic reaction with hydrogen to form ammonia (fully achieved by 1910) was to become a very significant step in the modernization (and purification) of chemical industry. Scaled up by Carl Bosch of Badische Anilin und Soda Fabrik, it meant that the prospect of mass starvation predicted by William Crookes in his Presidential Address to the British Association in 1898 was preventable—as of course Crookes had hoped. The flipside of Haber & Bosch's achievement was that in 1915 the German High Command (on the advice of Haber) would divert ammonia synthesis into nitric acid manufacture for the production of conventional explosives. The use of chemical weapons of warfare were prohibited by the Geneva Convention of 1908; hence one would not have predicted Haber's suggestion that the stalemate of trench warfare might be overcome by flushing British and French soldiers out of their dugouts by irritating them with chlorine gas. The escalation of this device into the production and use of more dangerous chemicals such as phosgene and mustard gas was to blacken the discipline of chemistry and begin the twentieth-century process that still affects chemistry in the twenty-first century—that chemicals are to be feared rather than trusted, and that DuPont's 1920s slogan, "better living with chemistry" became "better living *without* chemistry". Needless to say, such a sentiment is incompatible with civilization.

On the other hand, as opposed to the pursuit of physical chemistry as the future, there were chemists (usually traditionally educated organic chemists) who mistrusted physical chemistry and the use of physical methods in chemistry. None more so than Henry Armstrong who saw himself as chemistry's chief critic and gatekeeper. From today's perspective Armstrong's attitude often seems that of a buffoon. In the 1890s we see him orchestrating a campaign against the possibility that argon was a new element and prioritizing Dewar over Ramsay as President of the Chemical Society in 1897, and later Rayleigh over Ramsay and causing a good deal of disturbance within the chemical community of the 1890s. Nor did he like the idea of radioactive disintegration, and as for Arrhenius's dissociation theory, that was beyond the pale. Armstrong could see no future for chemistry unless it stuck to its disciplinary roots of analysis, atomic weight determinations, and structural determinations using chemical methods alone.

The phrase or by-line, "Chemistry of the Future", was first used by William Crookes in a report on Sir Benjamin Brodie's lecture on the Calculus of Chemical Operations given to the Chemical Society in 1867. Was Brodie's anti-atomic philosophy the future? What most struck Crookes was the calculus's implicit corollary that the elements must be compounds, thus reviving Prout's hypothesis that Jean-Servais Stas had recently rejected. Ten years later, in 1877, Crookes again used the phrase journalistically in his own *Quarterly Journal of Science*, referring this time to gaps in the periodic table and the possibility that the elements represented a past struggle for existence in their evolution as the basic material from which they had been created cooled down. The chemistry of the future lay in recognizing that elements were manufactured articles created by an evolutionary process and an investigation of the nature of the primary matter. It was this programme that led him to cathode rays or what he called the fourth, radiant, state of matter.

Curiously there was no use of the phrase "Chemistry of the Future" at the dawn of the twentieth century, despite H. G. Wells's penchant for technological predictions in the *Fortnightly Review* of 1901. As *Nature* commented at the time, it was easier to reconstruct a prehistoric animal than to predict the future, and the main use of prognostication would be to point to dissatisfactions with a present day situation. That is exactly the impression given by speeches at the British Association between 1901 and 1904. They all tended to be retrospective reviews rather than upbeat predictions of future glory, and gloomy prognostications about the failure of British governments to take scientific education and industry sufficiently seriously dominated discussion.

The same complaint was reiterated in speeches by James Dewar and the younger William Henry Perkin at meetings of the British Association for the Advancement of Science in next few years, as well as by Norman Lockyer in *Nature* and in the organization he founded in 1904, the British Science Guild. Despite British chemists' relatively tiny contribution to the world's chemical literature, it was obviously increasing at an exponential rate. That's why the Chemical Society introduced *Annual Reports* in 1905, the first volume covering the year 1904 and containing a review of radioactivity by Soddy.

It does seem that British chemistry was in the doldrums in the early 1900s. If one looks at Sieghard Neufeldt's *Chronologie Chemie 1800–1980*, there is really nothing of great significance published between 1901 and 1905 apart from William Pope's work on inorganic stereochemistry and Stanley Kipping's first evidence of a new class of compounds, the silicones. Compare this with the continent of Europe (mainly Germany), where Richard Willstätter had started his important synthesis of the tropines, Grignard had introduced organic magnesium reagents as synthetic intermediaries, Mikhail Tswett was inventing plant colours that would lead to chromatography, and new elements were still being discovered such as Eugène Demarçay's europium in 1901. But for anything more exciting and influential, one has to look at developments in biochemistry and radioactivity, and these sections are indeed the most important chapters of the first *Annual Reports*.

What then did British chemists living in the 1900s make of how chemistry should be performed in the future? As an example of Kopp's conceit, let's take the example of three leading chemists already mentioned who were at the peak of their careers when the elderly Queen Victoria died in 1901. What did William Crookes (1832–1919), William Ramsay (1852–1916) and Henry Edward Armstrong (1848–1937) make of chemistry and its likely future and of how it should be taught?

Crookes discovered in 1900 that uranium could be separated into an inactive uranium-X. This proved to be the key observation that pushed Ernest Rutherford and Frederick Soddy towards the model of radioactive disintegration. Crookes believed that the wonders revealed by his cathode ray investigations in the 1870s and 1880s, his work on radioactivity, and on phosphorescent spectroscopy all pointed in the direction of the unity of matter. It confirmed his admiration for historical figures such as William Prout (who first speculated that elements were complex), Michael Faraday (who had speculated the existence of a finer state of matter), and Jean-Servais Stas (who had brought analysis and the determination of atomic weights to great accuracy even though not believing in Prout's hypothesis).

Everyone remembers that Crookes got involved in psychic research in the 1870s. Less well-known is the fact that he joined Madame Blavatsky's Theosophical Society in 1883, probably because he genuinely admired its universalist concept that there was a kernel of truth in all the world's religions and that much could be learned from Eastern religions. In 1895 Annie Besant (who had studied some chemistry at London University), inspired by Crookes's work on radiant matter as the primary matter of the elements, developed a clairvoyant chemistry in which she and her partner Charles Leadbeater imagined themselves as Lilliputians peering into the interior of atoms and mapping their structures (Figure 6.1).

An elaborate version of this "occult chemistry" was published as a book in 1908. Was this the chemistry of the future? Crookes kept

Figure 6.1   Annie Besant (1847–1933), theosophist, birth control and women's rights activist and supporter of Indian self-rule. She studied chemistry at Birkbeck College but was refused a degree because of her atheism. Her clairvoyant 'occult chemistry' was developed with Charles Leadbeater in the 1890s. (Wellcome Images)

quiet and said nothing, though earlier in 1895 he had politely stated that Besant's models might offer clues on where to look for missing elements. "Occult chemistry" now looks very odd and suspicious and even a fraudulent pretense, but at the time it was no odder than what was happening in radioactivity and in the laboratory of William Ramsay at University College London (Figure 6.2).

Ramsay is best known for his work with Lord Rayleigh on the discovery of argon and his succeeding work in identifying the other rare or inert gases, including radon. His work with the latter led him into murky waters that are less well-known today, but which had an interesting postscript. In 1903 he and Soddy put radon in a spectrum tube and after a few days detected the presence of helium that had not been there previously. (This, of course, was due to the emission of alpha particles.) On the one hand this confirmed the disintegration theory Rutherford and Soddy had previously hinted at. On the other it led Ramsay into speculative experiments as to whether it might be possible to reverse radioactivity dissociation

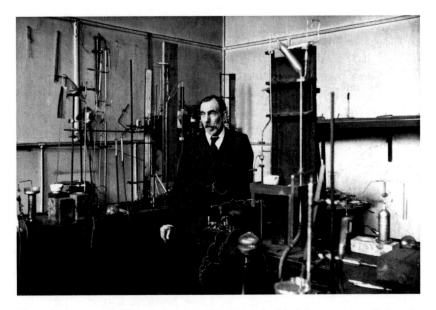

Figure 6.2   Sir William Ramsay (1852–1916) in his laboratory at University College London where he was professor of chemistry from 1887 to 1913. He was knighted in 1902 and was the first British chemist to be awarded the Nobel Prize in 1904 for his work on the inert gases. (Science & Society Picture Library)

(as he saw it) and to use its energy to synthesize elements. Beginning in 1907 Ramsay tried to transmute elements using radon as an energy source. He claimed that when pure water that had been exposed to radon was electrolysed, an excess of hydrogen was produced. And when the same was done with a copper salt he believed he detected lithium. Ramsay announced the transmutation of copper into lithium in a lecture to the London Institution in January 1907. Similar claims were made by him from then until 1912 when he announced that neon came from water and that he had synthesized argon from hydrogen and sulfur.

Although viewed sceptically by the majority of chemists and physicists (especially by Rutherford and Bertram Boltwood who had scathing things to say in their private correspondence about Ramsay's procedures—and use of stopcock grease), these announcements had a popular impact. Was transmutation possible? Had the old alchemists been right? What had the old alchemists been trying to do? In other words, it drove a few interested bystanders back into the history of chemistry.

So, the interesting conclusion is that Ramsay's curious work, like Crookes's, led to interest in chemistry's past. We are not surprised, then, to find that Ramsay's hero was the eighteenth-century pioneer of pneumatic chemistry, Joseph Black, whose life he wrote. Likewise, Henry Armstrong believed that the only wholesome future for chemistry was to return to the beliefs, commitments and training methods of the previous generation of chemists, whose lives he admired and upheld. Armstrong's future lay in simplifying the rudiments of chemistry and returning to its classical, historical roots rather than loading the tyro with the "jargonthropus" of physical chemistry which even he had to admit was transforming the frontiers of chemistry at the research front. As he said elsewhere: "On any mountain, the route of the historic first ascent is always invariably still the easiest route to the summit". In other words, the easiest approach to teaching and understanding a problem is by looking back to the first solution. Small wonder, then, that historians of chemistry find Armstrong, this quirky and outrageous Dr Johnson of chemists, a sympathetic character.

As Tennyson argued in his poem *In Memorium* in 1849, remembering our origins becomes a way of looking to the future. Crookes, fresh from warning the world that it would not have enough to eat unless nitrogen was fixed, believed that the wonders of radioactivity would transform the science. Ramsay (who was to win the Nobel Prize in 1904) used the inert gases that he, Rayleigh and Morris Travers had revealed to argue the necessity of applying physical methods in chemistry and that experiments with radio-active materials revealed the possibility of transmutation. Armstrong, who asserted that physical methods were leading chemists astray, stood up for organic chemistry as the focus of chemistry. Retrospectively, these individuals can be seen to have stood for the three dominant themes of twentieth-century chemical development and transformation: atomic structure (the electron); chemical physics (mathematical and instrumental methods); and synthesis (organic and macromolecules). All three also exemplified the links between chemistry and commerce and the importance of education in the laboratory.

The twentieth century turned out to be the century of the electron, the macromolecule and spectroscopy as well as the replacement of coal chemistry by oil chemistry. Little of this could have been foreseen by our three spokesmen, but each in his way looked to the past as a foundation for the future. Ironically, the chemist who really was to establish the future of chemistry, Linus Pauling, was born in 1901.

CHAPTER 7

The Alchemical Society, 1912–1915

Although several national chemical societies have historical sections or divisions devoted to their members' interest in the history of chemistry, there is only one international society that is dedicated to historical research in the subject. The Society for the Study of Alchemy and Early Chemistry was founded in November 1935 when a committee (later Council) was formed and advertised widely in the press during the following month. Although the society held no open meeting until one at University College London in November 1936, papers were solicited and accepted during the previous months of 1936 ready for the inaugural meeting and the launch of a journal entitled *Ambix* in May 1937.

In fact, The Society for the History of Alchemy and Chemistry (to give it its present title) was not the first such society in Great Britain. The Alchemical Society was founded in London at the end of 1912 by a mixed group of occultists, chemists and historians. In a letter published in *Chemical News* on 29 November 1912, a 25-year old Regent Street Polytechnic chemistry lecturer named Herbert Stanley Redgrove posited a need for historical investigations of the origins of the modern sciences. Because professional scientific societies like the Chemical Society were fully devoted to ongoing research in their respective disciplines, he suggested there should be other societies devoted to the history of every science, the most pressing need being one for the history of chemistry.

The Case of the Poisonous Socks: Tales from Chemistry
By William H. Brock
© William H. Brock, 2011
Published by the Royal Society of Chemistry, www.rsc.org

"At least, I think this is certainly the case with regard to Chemistry, since so much diversity of opinion exists as to the origin and nature of its parent Alchemy, and so many interesting questions arise in connection therewith. The alchemists wrote in a language, so it seems, entirely their own. Surely it had a meaning for them. Yet what [was] this meaning? Did their efforts to transmute base metals into gold arise from a belief that theological doctrines concerning the regeneration of the soul ought to be applicable to mineralogical problems? Or was it an outcome of the manufacture of artificial stones? Did the alchemists by their speculative methods grasp certain fundamental truths concerning the basic unity of matter, or were their theories utter folly unenlightened by any ray of truth?"

These were questions that needed answering by means of an alchemical research society.

But why form a society for the study of alchemy in 1912? There were two cultural triggers.

The first, as Alex Owen makes beautifully clear in her fascinating book, *The Place of Enchantment: British Occultism and the Culture of the Modern* (2004), is the way middle-class Victorians and Edwardians, faced by doubts of traditional religion, sought spiritual solace in spiritualism, theosophy and all sorts of esoteric clubs. This was an interest that was exploited by novelists such as Bulwer Lytton, Frances Barclay and Marie Corelli, whose novels dominated circulating libraries. Many of these movements, such as Blavatsky's theosophy, were a heady mixture of Western and Eastern religions, Neoplatonism, Boehmism and freemasonry. Collectively, they taught that there was a body of hidden knowledge (*i.e.* occult knowledge) that could be given to genuine adepts by a host of spiritual "masters" who lived on a different plane or dimension.

Alternatively, such occult knowledge might come from contemplating works of Hermetic philosophy, including alchemy. In 1884, Dr Anna Kingsford, a vegetarian and anti-vivisectionist Christian mystic, broke away from Blavatsky's Theosophical Society and founded her own separate Hermetic Theosophical Society which investigated the rich heritage of Western Christian mysticism rather than Eastern traditions. The same decade saw the formation of the Hermetic Order of the Golden Dawn which claimed direct descent from an earlier Rosicrucian order. The study of alchemy was one of the major elements in its higher-grade rituals. Two of its members were the American-born Arthur Waite (1857–1939), a prolific translator and editor of alchemical works and former copywriter for Horlicks malted milk drink; and the Liverpool artist Isabelle de Steiger (1836–1927), the widow of a wealthy Swiss merchant. It was Steiger who revived interest in Mrs Mary Anne Atwood's *Suggestive Inquiry into the Hermetic Mystery* (1850), which inaugurated the interpretation that alchemists had been seeking spiritual enlightenment not the physical transmutation of matter. Out of print since its publication, Steiger saw to its reprinting in 1918. She herself published *On a Gold Basis* with the firm of Philip Wellby in 1909, a strange book which Redgrove read as a newly graduated chemist. Waite dismissed it as unintelligible.

Following a sexual scandal in 1903, Waite and Steiger founded a rival Order of the Golden Dawn with a strongly Christian basis, leaving the old order to indulge in ritualistic magic. It is important to note that Waite and Steiger differed in their interpretation of alchemy. Waite believed the alchemists had been practical chemists, Steiger that they were religious contemplatives. One of the purposes of the Alchemical Society for Redgrove was to mediate between these two interpretations.

The second causal factor in the formation of the Alchemical Society, and the one that most excited Redgrove, was the fact that the recent study of radioactivity had made the possibility of elemental transmutations a rational possibility. Indeed, in *Nature* in 1907 William Ramsay had claimed to be able to degrade radon into the other rare gases and then into lithium by altering certain physical conditions. Although Ernest Rutherford rubbished these claims, Ramsay persisted in his experiments. In 1912 he reported to the Royal Society that he had created neon and helium from hydrogen under electrical discharge. His *Elements and Electrons,* also published in 1912, stood by such claims and also, intriguingly, speculated that the alpha particle would eventually turn out to be an energy bullet or Philosopher's Stone for the promotion of transmutation (in which he was proved right by Rutherford in 1919). To add to Ramsay's exciting claims, Frederick Soddy in 1912 recalled how Michael Faraday had once said that "transmutation is the final goal towards which chemistry should aspire". Soddy then added:

"The power to decompose and build up the known elements and to construct new ones at will as is now done for compounds, would elevate chemistry to an infinitely loftier plane than the rather secondary and subordinate position among the physical sciences it occupies at the moment."

In the same essay Soddy envisaged nuclear reactions as possibly supplying new sources of power—a speculation immediately taken up by H. G. Wells in his 1914 novel, *The World Set Free*.

The mission of the Alchemical Society was to "study the works and theories of the alchemists in all their aspects, philosophical, historical and scientific, and all matters relating thereto". The principal organizer was Redgrove who persuaded Steiger to join him and to recruit others such as Waite and Ralph Shirley, the editor of *The Occult Review*, and his publisher Philip S. Wellby. Shirley, the son of the professor of ecclesiastical history at Oxford, had founded *The Occult Review* in 1905, and devoted it to essays exploring the problems of life and death and the truths underlying all religious beliefs. Finally, Walter Old was recruited and the initial meeting held at the International Club in Regent Street in January 1913 when John Ferguson, professor of chemistry at the University of Glasgow and an eminent bibliographer of alchemical manuscripts and printed books, was elected Honorary President. There were three Honorary Vice-Presidents, the occultists Arthur Edward Waite and Isabelle de Steiger, both of whom contributed papers to the Society, and William Gorn Old, the astrologer, who also took on the secretaryship of the Society. Redgrove, who read a paper on "The Origins of Alchemy" at this inaugural meeting, agreed to produce and edit the Society's *Journal* which Wellby, a believer in the Atwood esoteric interpretation of alchemy who acted as the Society's Treasurer, arranged for the *Journal* to be published and distributed by the firm of William Rider & Co., a well-known occultist press owned by Shirley. William Rider & Co. specialized in books on the occult and the firm's publications embraced a special category labelled "alchemical philosophy" that included Waite's books on Paracelsus and his editions of alchemical classics such as the *Turba Philosophorum*. The firm's premises were destroyed during the Second World War, but the name has survived as an imprint of Random House. Redgrove also arranged for the Society's publications to be sold at the science and medicine bookshop owned by H. K. Lewis in Gower Street, next door to the chemistry department of University College.

The connection with Rider's was a good arrangement since it allowed the Society to conduct its affairs from the publisher's offices in Paternoster Row behind St Paul's Cathedral. Redgrove and Wellby ensured that the Society's meetings were well publicized and reported in *Chemical News, Nature, English Mechanic* and *Knowledge*, as well as in *The Westminster Gazette*. Clearly, the new society was accepted as an academically respectable learned institution. The Society planned to meet about eight times a year and charge a subscription of 10/6d [£1.05]. Within a year the Society was able to cease hiring a meeting room at the International Club and to meet free of charge at The Occult Club in Piccadilly Place, between Vine Street and Piccadilly. This was arranged by one of the Society's members, the spiritualist medium W. de Kerlov.

The membership seems to have consisted of roughly equal numbers of members of the esoteric and exoteric persuasion. A number of practising chemists joined, but not Ramsay and Soddy who were otherwise both interested in the history of chemistry. Perhaps they were suspicious of the Society's occultist membership and of its possible hidden agenda of re-spiritualizing science? A striking feature of the Society's membership, however, was the number of female members and that the society encouraged conviviality and social interaction by, firstly holding an annual dinner at Pinoh's Restaurant in Wardour Street in Soho; and secondly, by setting up study groups for "the mutual interchange of information and assistance to be effected by correspondence and informal meetings". Whether the latter arrangements were successful is not clear, but in any case, the Society collapsed before this could have developed very far.

Another interesting feature was that the Society exchanged information with its French equivalent, La Société Alchimique de France (founded in 1896) and elected its President, the practising alchemist François Jollivet-Castelot as an Honorary Member. In their turn, Ferguson and Redgrove were elected to the French society which also made its journal, *Les nouveaux horizons de la science et de la pensée* (1906–1914), available to the British membership. Redgrove, whose parents managed a large boarding house for male and female servants in Tottenham Court Road, was an efficient editor; he produced three volumes of the Society's *Journal* between 1913 and 1915. Many of the articles are still worth reading. Each volume was of some 77 pages and priced 1/- to members and 2/- to non-members, sales being conducted through Rider's and Lewis's bookshops.

Knowing that the annual subscription was 10/6, we can deduce from the first published balance sheet that the initial membership was about 46. This rose to about 50 in 1914, but dropped sharply to 37 in 1915 when the war shattered any hopes that the Society would thrive. At a meeting on 8 October 1915 it was agreed that meetings would have to cease for the duration of the war. Redgrove produced the final volume of the Society's *Journal* at the end of 1915.

Redgrove, who held a London University BSc (1907), is best known for his book *Alchemy Ancient and Modern*. First published by Rider in 1911, it was revised in 1920 and re-edited by Harry Sheppard, an active member of the Society for the History of Alchemy and Chemistry, in 1973. Sheppard noted that Redgrove's interpretation of alchemy had been largely endorsed by later historians, namely that alchemists had often used the language of Christian mystics and Greek philosophy and that, in principle, they had been correct in believing in the possibility of transmutation, though the vast quantities of energy required to effect elementary transmutations rendered it a waste of time. Redgrove, a friend of the schoolmaster and rhenium specialist Gerard Druce, had earlier published a book on the calculation of thermochemical constants and another on Christian mysticism, *Matter, Spirit and the Cosmos* (1910), which was strongly anti-materialistic and much influenced by the reading of Emanuel Swedenberg. Parts of this book had already appeared as essays in Shirley's *Occult Review*. Redgrove, who accepted some aspects of spiritualism and telepathy, had been struck by ideas of the fourth dimension and concluded idealistically and opaquely that the solution to the riddle of existence was:

"Infinite Love, the final cause, demands an object of affection [*i.e.* man]. Infinite Wisdom supplies the means, and Infinite Power brings into being that which Infinite Love desires."

It is not known what Redgrove did during the First World War which saw the loss of several members, including the Society's second secretary and prolific contributor, Sijil Abdul-Ali, who was killed in France in 1917 at the age of 28. The Society's President, John Ferguson, died in Glasgow in 1916. William Ramsay had died of throat cancer in July 1916, by which time it had become clear that his strange experiments had been faulty and downright careless. In one "transmutation", for example, lithium had been "formed" from Ramsay's cigarette ash.

Nevertheless, Redgrove hoped to restart the Society after the war with a wider remit to include historical study of the Kabbalah, astrology and talismanic and ceremonial magic, but it did not materialize. Nor were German scholars any more successful. In 1927 Otto Wilhelm Barth, who had inherited the famous science publishing house of J. A. Barth in Leipzig, founded the journal *Alchemistische Blätter*. He soon used this to announce the formation of a German Alchemical Society with groups that would meet in Berlin and Hamburg. Although the German journal attracted a few articles of an historical nature, most of the contents were devoted to the esoteric uses and interpretation of alchemy and the ways it might be used for the reformation of German economy and culture. After sporadic publication the journal and its society appear to have collapsed in 1930.

Redgrove himself moved from technical college instruction into school teaching at Battersea Grammar School in the 1920s (where he was a colleague of Druce's) as well into socialism. At the end of the 1920s Redgrove performed his own self-transmutation by taking up chemical research, devoting himself to the scientific improvement of women's hair dyes and beauty products and setting up Berkshire Beauty Products with his wife. He published the delightful *The Cream of Beauty* in 1931. Although he continued to contribute to *The Occult Review*, his private interests had clearly moved away from alchemy to judge from the fact that in 1940 he published a book on the history of airmail postal delivery services.

Part 2: Organizing Chemistry

The traditional professions were restricted to the law, the church and medicine. But with modernization (as sociologists describe the processes whereby social stratification became more determined by intellectual knowledge and physical skills) new professions emerged. Among the new career paths was that of science, the term "scientist" being first employed to describe the profession in 1834. This was the period when the newly formed British Association for the Advancement of Science first began to meet in recognized, specialized sections devoted to mathematics, chemistry, engineering, and so on. The eight stories in this section deal, in various ways, with the way chemistry became organized as a discipline in the nineteenth century. Scientific disciplines require an educational system that allows and supports specialization, professional and learned societies, agreement upon a common language and the standardization of units throughout the subject and, in the case of a practical subject like chemistry, laboratories. The opening essay provides an overview of the development of the British educational system in the nineteenth century and of how, by the 1870s, educational facilities at secondary school level encouraged the emergence of a scientific community well-grounded in chemistry.

Although no national chemical society emerged until 1841, the second story illustrates how attempts to form a society were made some two decades before by skilled artisans who were suspicious of the financially independent "gentlemen" natural philosophers like Humphry Davy and William Hyde Wollaston who, until the mid-1830s, professed their interest in chemistry through the elitist Royal Society. The 1840s saw not only the foundation of the (Royal) Chemical Society and (Royal) Pharmaceutical Society, but also the Royal College of Chemistry (RCC) in Oxford Street. The latter, under the genial leadership of A. W. Hofmann, transformed chemistry teaching and research and provided analytical man-power for Britain's existing and future chemical industries. An interesting contemporary view of the state of that industry is provided by a critical German visitor in the third story.

We stay with laboratories in the following two essays. The RCC was modelled on the great teaching laboratory that Liebig had founded in Giessen in 1824, and although he was not entirely responsible for the creation of the modern chemical laboratory, Liebig was undoubtedly a powerful influence in making it an iconic feature of European chemists' training and in the promotion of chemical research. Today we take it for granted that the other sciences also require laboratories. It was not always so. In fact, it was the prior model of the chemical laboratory that promoted the creation of physics laboratories and the teaching of practical physics. It was the scarcity of academic jobs for chemists that encouraged some of them to use their skills in the promotion of physics teaching.

One of the particular features of chemistry is that, like mathematics, it possesses its own private language of symbols and formulae. It is this feature that makes the popularization of chemistry among laymen so difficult to the extent that chemistry becomes under-appreciated and its successes in bettering society fail to gain the recognition which astronomers, physicists and biologists seem to gain effortlessly. But the key to modern chemistry's position as the "central science" of today undoubt-edly lies in this private language and the ability of chemists to manipulate molecules virtually on paper. As the sixth essay on "chemical algebra" shows, however, systematization and stan-dardization were only achieved with difficulty. Despite their use of a private language, chemists are still sociable beings—and alcohol has been one of chemists' favoured molecules. The penultimate story pays tribute to the Chemical Society's "B-club" and shows chemists outside their laboratories gossiping and feasting and inventing drinking songs that mercilessly debunked their con-temporaries. Such droll activities still form part of chemists' culture. They cement lifelong friendships and probably even research collaborations. They are part of professional training and the tuition of the following generation. A final brief story marks a return to the laboratory and the question of the role that it plays in the making of the chemist.

CHAPTER 8

Putting the 'S' in the 'Three R's'

"Scientific education," wrote the young chemical assayer, William Stanley Jevons, in 1856, "is one of the best things possible for any man, and worth any amount of Latin and Greek. It tends to give your opinions a sort of certainty, force and clearness which form an excellent foundation for other sorts of knowledge less precisely determined and established, provided that you do not let your mind become completely formed to science". Indeed, Jevons did not let science dominate his life, for he became lost to chemistry and a well-known Victorian economist instead. His allusion to the "worth" or value of science was probably a reference to a famous and internationally influential essay on education, "The knowledge that is most worth", which had been published by Herbert Spencer two years earlier in 1854. As a spokesman for a new industrialized world (symbolized most recently by the Great Exhibition) that cast aside the classical training of young men and the ornamental learning of young women, Spencer had rhetorically raised the question, "What knowledge is most worth?" To this question, he said, there is only one answer, "Science".

> "For direct self-preservation, or the maintenance of life and health, the all-important knowledge is—Science. For that indirect self preservation which we call gaining a livelihood, the knowledge of greatest value is—Science. For the due discharge of parental functions, the proper guidance is to be

The Case of the Poisonous Socks: Tales from Chemistry
By William H. Brock
© William H. Brock, 2011
Published by the Royal Society of Chemistry, www.rsc.org

found only in—Science. For that interpretation of national
life, past and present, without which the citizen cannot rightly
regulate his conduct, the indispensable key is—Science. Alike
for the most perfect production and present enjoyment of art
in all its forms, the needful preparation is still—Science; and
for the purposes of discipline intellectual, moral, religious—
the most effective study is, once more—Science."

From the remarks of Jevons and Spencer, then, it would seem
likely that the 1850s was the decade in which the need for science in
the educational curriculum of English schools was being pressed
most strongly: the "S" of Science was to be added to the traditional
curriculum of the "3R's" of "Reading, 'Riting and 'Rithmatic".
[The phrase is said to have been first used in a toast proposed by the
plebeian Lord Mayor of London, Sir William Curtis (1752–1829)
in the 1820s.] Of course, reading, writing and arithmetic were not
knowledge as such, but instruments for the gaining of knowledge.
The trouble with this, according to the spokesmen of science in the
1850s, was that the vast majority of Victorian Britons got no further than these primary instruments.

By way of generalization and simplification, it can be suggested
that there were two phases or waves of enthusiasm for introducing
science teaching into English schools. These roughly correspond to
what we call today elementary and secondary education.

THE FIRST WAVE: SCIENCE IN ELEMENTARY TEACHING

Until the 1830s the provision of elementary education for working
class children was the responsibility of voluntary agencies, the
Anglican-based National Society for Promoting the Education of
the Poor in the Principles of the Established Church (founded
1811), and the largely Nonconformist-based British & Foreign
School Society (founded 1814). To all intents and purposes,
instruction in such "church" schools was restricted to the 3R's.
However, following the arrival of government aid to such schools
in the 1830s, there were some interesting and influential experiments at introducing young children up to the age of twelve to the
rudiments of science. Three examples can be mentioned—those of
Richard Dawes (1793–1867), John Henslow (1796–1861) and
David Boswell Reid (1805–1863).

The Reverend Richard Dawes was a Yorkshireman who read mathematics at Trinity College, Cambridge, under the omniscient William Whewell. He was a Fellow of Downing College, Cambridge, for some years before he took Holy Orders in 1836. In 1842, taking advantage of the system of government grants for National Schools, Dawes opened a school in his parish at King's Somborne, an agricultural village in Hampshire. Here he insisted on emphasizing the reading of *secular* literature of a mainly scientific and geographical kind with a view to giving children information that was explicitly concerned with "the concerns of everyday life". He did not of course ignore religious instruction, but the books used in the school were those hitherto published by the government for exclusive use in Irish schools where, because the majority of the population was Roman Catholic, non-sectarian textbooks had had to be introduced. More particularly, Dawes articulated and developed what became known as "object lessons", in which a topic (such as "bread" or "coal") was used by the teacher to weave information of a practical nature that ranged over a variety of disciplines and fields of knowledge.

Dawes's work impressed one of the government's school inspectors, the Reverend Henry Moseley (1801–1872). It was he who widely diffused Dawes's ideas and practices through the Committee of Council on Education and its publications. Moseley himself [the father of the naturalist, Henry Nottidge Moseley (1844–1891), and grandfather of the physicist, H. G. J. Moseley (1887–1915)] ended his career as a canon of Bristol Cathedral, where he helped create a Trade School for the sons of the upper working class. This became the distinguished technical college known, after 1885, as the Merchant Venturers' College, its building being designed by E. C. Robins (mentioned below).

If it had not been for the emergence of the Department of Science and Art in the 1850s, which came to see the proper place of scientific instruction in secondary, not elementary, education, Victorian England, Wales and Ireland might well have seen science teaching formally established within the state's elementary schools. In practice, however, with the advent of the Revised Code of the elementary curriculum in 1862, state elementary schools became confined to instruction in the 3R's only. It transpired, therefore, that Dawes's exploitation of object lessons, and the suggestions expressed in his book, *Hints on an Improved System of National Education* (1847), became much more influential within secondary, rather than elementary, instruction. It is possible to trace Dawes's influence on, for example, the object-type teaching devised by the physicist Frederick Guthrie in the 1870s, and on the science of the environment which T. H. Huxley called "physiography" and which became the "general science" taught in many twentieth-century schools. Ironically, it is only since the establishment of a "national curriculum" in the last few decades that junior (elementary) schools in Britain have been obliged to introduce children to science before the age of eleven.

We should note in passing that King's Somborne was geographically adjacent to an extraordinary Quaker secondary school called Queenwood College, formerly the Harmony Hall of Robert Owen's unsuccessful utopian communal experiment. Dawes was well aware of the quality and amount of science teaching conducted at this private school for farmers' sons and young men who intended to make careers in civil engineering. Its teachers in the late 1840s, and with whom Dawes rubbed shoulders, included John Tyndall and Edward Frankland who, following scientific studies at the University of Marburg, were soon to launch themselves on dazzlingly successful careers in physics and chemistry. Both men were to join Spencer in the 1850s and 1860s in clamouring for science teaching in the public schools and the new secondary proprietary schools. They fully agreed with Dawes that science was useful knowledge and something that all young people could benefit from in a practical sense in their adult lives.

A rather different rationale for teaching science to elementary school children was that of another Anglican priest, the Reverend John Henslow. He is, of course, best known as Professor of Botany at the University of Cambridge, where he inspired and befriended Charles Darwin. But he was also the parish priest at the village of Hitcham in Suffolk and it was here that he taught village schoolchildren botany in the National School.

Henslow, like many members of the scientific community of his generation, was inspired by faculty psychology—put simply, the concept that the mind comprised a number of separate organs, or faculties, such as observation, reasoning, verbalizing, *etc.* Henslow believed, like many of his generation, that the traditional 3R's, let alone traditional university teaching, failed to adequately stimulate and train important areas of the mind. He was convinced that science-trained powers of observation, memory and reasoning that were largely left untouched by reading, writing and basic arithmetic. The most favoured elementary science to achieve these goals was, according to Henslow, botany—especially the observational and taxonomic system of plant identification, naming and classification. From today's perspective, Henslow appears to have had a rather Gradgrindish attitude towards the children he was educating with his insistence on their learning the correct technical names for the parts of flowers and their Latin names; on the other hand, we know from our experience of modern children that they love pronouncing and spelling the jaw-breaking names of dinosaurs!

Like Dawes, Henslow's influence was to be on secondary education, as passed to and transformed by the work of the Reverend James Wilson at Rugby School in the 1860s. But before examining this, we should mention a third approach and influence on later school science education (Figure 8.1).

The third approach to the introduction of science into elementary teaching came from a Scots chemist, David Boswell Reid, who demonstrated the possibility of installing laboratories in schools.

Figure 8.1   James Maurice Wilson (1836–1931), clergyman, mathematician and pioneer science teacher at Rugby School. He later became headmaster of Clifton College where he established a first-class science teaching staff. (Brock)

During the period when Reid was an extra-mural chemistry lecturer in Edinburgh, he designed and had built a chemistry laboratory catering (reputedly) for a hundred students. In this workshop, an illustration of which survives, he developed a flat-glass technique of microanalysis, or spot-testing, as well as instruction in preparative chemistry. On coming to London in 1834 armed with a scheme for ventilating the new Houses of Parliament (the former Palace had burned down in 1834), he introduced his microanalytical methods into a few London elementary schools with the approval of the Committee of Council on Education. It would seem that Reid's pedagogic philosophy was that working class children needed to know about ventilation, diet and sanitation if they were to cope successfully as adults with the appalling social and urban conditions of the period, and that chemical knowledge was the tool they most needed.

Each and all of these three approaches—useful knowledge, the training of observational and logical powers, and the honing of practical laboratory skills—also embraced the conviction that science was a way of improving the moral and religious convictions of the young. As James Wilson put it later, science conveyed feelings of wonder and awe that complemented or induced religious convictions. Despite this, the 1862 Revised Code effectively closed the door on science in the primary school. It was only much later in the century that the Department of Science and Art's examination system allowed science back into the state system *via* the back door of the so-called Higher Grade School. These were elementary schools that gained the Department's grant aid by offering science examinations to pupils over the age of twelve. It was this "anomaly" that was to lead to the famous legal case in 1897 (the Cockerton judgement), which effectively brought about the 1902 Secondary Education Act.

Since 1860, however, there had been an effective change of emphasis from elementary to secondary science teaching. The reason for the change of emphasis is clear. The schemes of Dawes, Henslow and Reid had evolved *ad hoc*, without the support and interest of the emerging scientific community. As that community emerged with growing strength after 1851, its members seem unanimously to have taken the view that science was best taught after thorough basic education had been achieved and that the physical sciences were better suited to such instruction. This was

Lyon Playfair's view, for example. And as a chemist and a pupil of the German, Justus Liebig, whose Giessen laboratory was so influential in training of British chemists, it was chemistry that made the running in the period 1860–1890. Moreover, since Playfair was closely associated with the early years of the Department of Science and Art (DSA), that government department was able to ensure that finance for science went towards secondary and technical education (the two categories were virtually identical in the DSA context) and so left the Education Department to finance the teaching of the basic 3R's in elementary schools. This policy had the approval of the scientific community which could, in any case, supplement its own precarious individual living from scientific teaching and research through the setting and marking of government examination papers. Moreover, the scientists' organizations—the Royal Society, the Chemical Society and the British Association—began in the 1860s to push hard for the introduction of science in endowed and proprietary schools to which their own middle class sent their sons. It will be noted, too, that these same schools provided welcome jobs and positions for many members of the scientific community.

THE SECOND WAVE: SCIENCE IN SECONDARY SCHOOLS

Ornithopachynsipaideia is not a word in the *Oxford English Dictionary*. It was a jaw-breaking term coined by James Wilson (1836–1931) when he was a young science master at Rugby School in the 1860s. He used the word to describe the way children were frequently crammed with subjects like fattened birds with pre-masticated foods. Wilson believed that children would cope better with their school subjects if one followed the other successively in a stratified sequence. "Bifurcation," or the study of two or more science subjects, might then follow in the upper school (from the age of sixteen onwards) when the relationship between subjects could be appreciated and specialization might begin. Before then, the purpose of all elementary science teaching (and he meant by this the rudiments of botany, natural philosophy and chemistry) should follow the principles laid down by Henslow and Dawes—to stimulate a thirst for knowledge, to fire the imagination, to open the eyes and ears to the objects and interests of the sciences, and to provide a solid body of factual information upon which the later, more specialized, study might crystallize. In contrast to Reid, however, Wilson laid little emphasis upon the "hands on" experimental method for children until the preliminary information stage had passed; on the other hand, to satisfy a child's natural curiosity and wonder, it was desirable that some experiments and demonstrations should be performed both by the teacher and by the pupils. This was the origin of Rugby's first laboratory in 1860 in a converted town hall cloakroom. The following year, Wilson's headmaster, the Reverend Frederick Temple, caused a lecture room and laboratory to be built in the school grounds; this was extended to hold thirty pupils when William Butterfield designed extensions to the school. By then, the whole middle school received two hours of science teaching a week, and Wilson's work became deservedly well-known through the evidence he gave to the Clarendon Commission in 1864. His outspoken comments on the drawbacks of teaching geometry with Euclid's text also made him something of a hero among members of the scientific community. Wilson further enhanced his reputation when, as headmaster of Clifton College from 1879 to 1890, he employed and encouraged a brilliant team of science teachers, several of whom were Fellows of the Royal Society.

It was Wilson who diagnosed acutely the practical difficulties and objections to science teaching in English public schools: the supposedly overcrowded classics curriculum that left no space for new subjects like science; the expense of laboratory teaching; the lack of trained science teachers; and above all, the lack of science scholarships at Oxford and Cambridge which discouraged schools from offering scientific tuition. Through modest example and the support of a sympathetic headmaster (Temple) and the scientific community, Wilson was able to transform the situation. In a long lifetime (he died only in 1931), this "Nestor of Science Teachers" lived to see the public schools—through their powerful Association of Science Masters founded in 1901—lead the field in curriculum innovation and become exemplars for science teaching in state secondary schools following the 1902 Secondary Education Act.

By then, largely through the influence of two chemists, Edward Frankland and Henry Edward Armstrong, school laboratories had become an obligatory feature of secondary school design and architecture. Frankland's efforts were achieved through the Department of Science and Art after he had complained as an examiner that students' examination answers showed an appalling ignorance of practical chemistry. Through his influence as chief examiner, the Department of Science and Art began in the 1870s to give building grants for laboratories. The designs of these laboratories owed much to a former science teacher at the Woolwich Military Academy, Captain William de W. Abney. When the Department of Science and Art reached the end of its life in 1902, Abney proudly noted that there were 1,165 registered laboratories for the teaching of chemistry, physics, biology and mechanics compared with a mere 65 in 1877. It has been further estimated that by 1900 there were about a thousand science graduates teaching science in British schools. Not surprisingly, given the window of commercial opportunity presented by this expansion, many architects began to specialize in laboratory provision. The most important of these was Edward Cookworthy Robins, whose *Technical School and College Buildings* (1887) is a superb source of illustrated information on both British and continental laboratory design.

Abney's name and military title reminds us, finally, that science teaching was significantly influenced by the reform of the British army in the wake of the Crimean War and the Indian Mutiny. On the one hand, many cashiered army personnel, like Abney, found their way into administrative roles in the Department of Science and Art where their scientific knowledge, technical expertise and organizing abilities were prominent. Even more significant, however, was the fact that the introduction of military examinations after 1857 for entry to Sandhurst and Woolwich (a process greatly encouraged by Prince Albert) forced public schools under parental pressure to introduce science teaching in their "modern sides". It was in this manner, as Prince Albert had hoped, that science became an acceptable element of general and gentlemanly education, as well as for the needs of professional military education.

In conclusion, beginning with a number of experiments in the 1840s and 1850s in introducing science in elementary schooling, under the influence of the needs of military educational reforms, technical education under the auspices of the Department of Science and Art, the practical example of teachers like Wilson, and through the forceful rhetoric of spokesmen for science such as Spencer, the Victorian era ended with science firmly entrenched as a necessary part of the secondary educational curriculum. We might go further and suggest that Britain ended the century better endowed with school laboratory facilities than anywhere else in the world.

CHAPTER 9

The London Chemical Society, 1824

In their standard history of the Chemical Society (now the Royal Chemical Society), which was founded in 1841, T. S. Moore and J. C. Philip noted that "an association of persons interested in the advancement of chemistry was ... nothing new, for in the preceding decades various chemical societies, generally of a local character and in most cases short-lived, had been started". Amongst these were: the Lunar Society's London-branch chemical society which met at the Chapter Coffee House in Paternoster Row during the 1780s; various Scottish and American university societies of roughly the same decade; the Royal Society's assistant society, the Animal Chemistry Club (1808–1825); and the London Chemical Society of 1824, the subject of this essay. According to Moore and Philip:

> "The London Chemical Society ... was founded in 1824 and had Dr Birkbeck for its first and only President. Its early demise was possibly not unconnected with the criticism which appeared in the columns of the *Chemist* (4 December 1824) to the effect that the lecturers of the Society were 'not sufficiently pregnant with the matter to give information in an intelligible manner'."

Two points emerge from this account. First, there is an implication here that the Society was created by George Birkbeck (1776–1841), the major force, if not the creator, of the Mechanics'

The Case of the Poisonous Socks: Tales from Chemistry
By William H. Brock
© William H. Brock, 2011
Published by the Royal Society of Chemistry, www.rsc.org

Institute movement. However, in 1824 Birkbeck was at the height of his fame, an extremely busy medical man who devoted all of his spare time to the vociferous demands for technical education of the emerging mechanics' movement. Despite Birkbeck's ubiquitous nature it seems unlikely that he would have created the society by himself. It is surely significant that Robert Warington Jr pointed out that one reason for the foundation of the Chemical Society in 1841 was that his father had the time and the freedom to engage in the necessary organization. It is also worth noting that Birkbeck was only asked to help create the London Mechanics' Institute after the idea had been mooted by others.

Second, there may be an assumption in Moore and Philip's statement that the 1824 Chemical Society was a gathering of "educated" upper and middle class chemical philosophers. The true situation was rather different, as a closer study of the journal *The Chemist*, reveals. But Birkbeck's name should provide the clue: for, in fact, the Society was yet another manifestation of the Mechanics' Institute movement.

The Chemist was a weekly journal, price three pence, which appeared in two small volumes between 13 March 1824 and 16 April 1825. Its publishers, John Knight and Henry Lacey, also produced the more famous and more successful *Mechanics' Magazine*, and like the latter, it was aimed at the artisan classes. Both journals were printed on poor quality paper, with crude typography and illustrations. The editor, who always maintained a strict anonymity, was of a strongly practical and anti-theoretical bias, as may be judged from his fulminations against time wasted by "academic" chemists in atomic weight determinations. The editor was, in fact, the political radical Thomas Hodgkin (1787–1869), who also edited the weekly *Mechanics' Magazine* which he founded in 1824.

On 22 May 1824 an anonymous correspondent, "A.W." (later amended to "A.M.") suggested, with editorial approval, the formation of "a nucleus for a society of young chemists, who might, at their common expense, purchase chemical texts, instruments, &c. as they want them, and that without bringing down [financial] ruin on any of them". The society was to consist of select respectable members and to meet several times a week. This suggestion for a kind of chemical mutual improvement society caught the imagination and approval of *The Chemist*'s readers, several of whom offered further support and suggestions for the venture.

To find a published discussion of how a scientific society should be set up and run is unusual though not unique. One correspondent, "J.G.", thought that membership should be restricted to about twenty people, that the subscription set at a shilling per month, and that ballots should be held to decide which books and apparatus might be purchased. The same writer, who seems to have been wise to the problems of adult education, foresaw some difficulties, and how they might be overcome:

"Supposing a Society to be formed, and to be furnished with a moderate apparatus and library, the next thing to be considered is how are the members to study? They cannot all meet, and each set about some experiment according to his own fancy, for then everything would be confusion. If the Society is to be continued, the business of it must be conducted with gravity and decorum. My opinion is, that the Society should take an experimental system of chemistry, Henry's, for instance, and go regularly through it; performing all the experiments contained therein, as far as their skill and apparatus would permit; and that every member, in his turn, should give a LECTURE, illustrative of the experiments performed."

By following a textbook such as William Henry's *Elements of Experimental Chemistry* (9th edn, 1823) the member would avoid the danger of nescience, the blind leading the blind. After continual practice, each member's lecturing and experimental ability would improve so that "amusement and instruction would certainly be obtained by every member at every meeting". It seems possible that some such practice was followed in the constituted society.

A more immediately practical proposal was made on 29 May 1824 by William Jones, who offered the use of a room, his own library and his apparatus for a period of one year. At the same time another correspondent, "Luzitanus", suggested that a meeting of interested people should be called immediately. He offered five guineas to help float the society, as well as the use of "all the works of chemistry that my library contains, until the funds will admit of the purchase of them for the Society: amongst them I have Parkes, Accum, Mackenzie, Chaptal, Cadet, Herpin, &c". Luzitanus failed, however, to offer hospitality for an inaugural meeting.

On 19 June 1824, "A.M.", the original proponent of the society, further elaborated his own ideas. It would be unfair to permit

Luzitanus to subscribe five guineas unless everyone were to do so; but this was too large a sum for most people. J.G.'s suggestion that the subscription should be a shilling per month was, on the other hand, too low, and A.M. proposed instead a shilling foundation fee and a similar sum every week. A.M., however, did not want to get involved in the government of the society—a sentiment shared by *The Chemist*'s editor who, however, promised every support for the society through his journal.

By June 1824, some "sixteen gentlemen" had expressed sufficient interest in the mutual improvement association to justify the editor's recommendation that they should all take up Jones's offer and meet at his rooms at 55 Great Prescot Street, close to the Royal Mint. Jones himself did not attend the first meeting on 26 June, though he reported a fortnight later that because the numbers attending had been few, a better publicized meeting would be held on 15 July. At this much more successful gathering Jones was elected temporary secretary of a small committee which was charged to frame the new society's rules. The three other committee members were Messrs A. J. F. Marreco (chairman), H. Fenner and J. B. Austin. Marreco was, of course, the previously anonymous "A.M."

A month later, on 12 August, the "London Chemical Society" was formally launched for "the study of chemistry in all its branches" and constituted as follows. There were to be two classes of membership, honorary and ordinary. The government of the society was to be in the hands of a president and two vice-presidents, a treasurer, a secretary, a curator, and a council of five. All officials were to be elected annually and they were charged to hold business meeting every fortnight. The election fee for ordinary members was one guinea and the annual subscription was two guineas payable in two parts half-yearly.

"The lecture room and library of the Society are to be open from nine in the morning until ten in the evening, except Saturdays, when it is closed at three o'clock. The ordinary meetings of the Society are to be occupied by the delivery of lectures in various branches of chemistry, by the reading of memoirs, by the performance of experiments, or by the discussion of any subjects connected with chemistry, as the council may direct and approve of."

No further information on the Society's activities is to be found until October 1824 when *The Chemist* reported that there had been an increase in membership, most of whom were young men.

"On Thursday, October 7th, we attended a Lecture on Heat, delivered by a Mr Davis [later corrected to Mr J. B. Austin] in a very pleasing style, agreeably illustrated by experiments. Several ladies were present taking warm [*sic*] interest in all that was said, encouraging the lecturer by their smiles."

The women present were guests, and not members, for in his presidential address Birkbeck revealed that women were not eligible for membership. It will be recalled that the present Royal Society of Chemistry only admitted women to the Fellowship in 1920 after a long struggle.

Lectures and meetings were held in hired rooms at 18 Aldermanbury, near the City of London Guildhall. Besides Austin, the names of only one other lecturer are recorded. This was Charles Frederick Partington (d. 1857), a professor at the London Institution at Finsbury Circus, a professional lecturer and author of several semi-popular books on science.

At a general meeting held on 18 November 1824 elections were held and filled as follows.

President: G. Birkbeck   *Vice-Presidents:* J. F. Cooper
                                                    A. J. F. Marreco

Treasurer: G. Smith   *Secretary:* W. Jones

Curator: J. B. Austin

Council: T. Dell, C. Dundersdale, H. Fenner, J. J. Silva, W. S. Stratford

Apart from Birkbeck, only three other names are known to historians. Antonio Jonquin Freire Marreco was a Portuguese wine merchant in the City. He subsequently moved to Newcastle, where he became a banker and director of the Stanhope & Tyne Railway. His son, Algernon Freire-Marreco (1836–1882) became a Reader in Chemistry at the medical school in Newcastle, as well as a chemical consultant. He helped to found the Newcastle Chemical Society in 1868 and later he became a founder member of the Institute of Chemistry.

One other member cited in *The Chemist* was Augustus Mongredien (1807–1888), the seventeen-year old corn merchant who in later life became a writer on free trade.

Misprints of initials were common in *The Chemist* and it seems probable that the vice-president, J. F. Cooper, was John Thomas Cooper (1790–1854). He was an ingenious chemical inventor and manufacturer, and lecturer in chemistry at Mount Street, Lambeth, the Russell Institution near Blackfriars Bridge, the London Mechanics' Institute and the Aldersgate Medical School. Later, when his pupil Robert Warington founded the 1841 Chemical Society, he became one of its more energetic early Fellows. The only other member of Council known to have joined the later society is George Smith, who worked at a brewery in Whitechapel.

Birkbeck, "with his usual zeal for the diffusion of knowledge", had willingly lent his name to the Society's enterprise just as he had to the London Mechanics' Institute the year before. On the 25 September 1824 he officially inaugurated the society at the City of London Tavern in Bishopsgate Street before an audience of an estimated three hundred people. At this meeting Jones reported that the Society:

"… have [*sic*] received communications, accompanied by donations of books from various authors of eminence. From the prosperous state of the Society, and the great increase latterly in its members, they sincerely hope that the Council, at the next Annual meeting will have communications to lay before it worthy of the Society."

In his address, on the theme of change and chemistry, Birkbeck was equally optimistic and enthusiastic. Since the purpose of chemistry was to "discover and explain the changes that occur amongst the integrant and constituent particles of different bodies", and to exploit this knowledge both to increase human material comforts and to demonstrate the divine design of the universe, cooperation of like minds in enterprises like the London Chemical Society would be of the greatest benefit and importance. For chemistry, he continued, was not merely of interest and benefit to learned and educated men, nor was it to be confined to men only. This important science could only develop rapidly and efficiently if knowledge of it were diffused throughout mankind, learned and unlearned alike. In Birkbeck's view, then, the Society was complementary to the aims of the Mechanics' Institute and to the contemporary movement for the Diffusion of Useful Knowledge.

Jones's and Birkbeck's optimism proved a damp squib for the Chemical Society proved a dismal failure. Although *The Chemist* continued publication for a further six months, no further reports of the Society's activities were reported. Indeed, beyond a small report of Birkbeck's lecture in *The Gentleman's Magazine*, no other contemporary periodical—not even *Mechanics' Magazine*—appears to have noted its existence. The fact that important organs of publicity, and chemical interest such as *Annals of Philosophy* and *Philosophical Magazine* ignored it is a reflection on the divide between the chemical natural philosophers and the professional chemists such as Partington and Cooper who staffed the country's breweries, dye-works and emerging gasworks.

It is doubtful whether the Society lasted into 1825. There may have been disagreements among the Council members and financial troubles—both were common difficulties in the contemporary Mechanics' Institutes. The standard of lecturing may well have been poor amongst a group of tyros, and as the Mechanics' Institutes discovered, professional lecturers of the calibre of Partington and Cooper commanded high fees. Like Hodgkin, many of the members appear to have been political radicals; consequently, they may have found political activism outweighed their enthusiasm for chemistry. Finally, it seems clear that the very nature of the Society would have militated against its continued success. Apart from Cooper, there were no learned chemical thinkers and experimentalists among them. The blind were leading the blind. Either through ignorance or intellectual arrogance, the principal London philosophical chemists such as Brande, Davy, Faraday, Phillips, Prout and Wollaston chose not to rub shoulders with the Society's rude, would-be chemical philosophers. By the same token, *The Chemist* failed as a commercial venture, as all similar ventures failed until William Crookes found a successful formula in the weekly *Chemical News* at the end of 1859. One undoubted reason for the success of the 1841 Chemical Society of London was that Robert Warington, its founder, ensured that it would be a fruitful amalgamation of the technological and academic chemist—even though this rapport only lasted until the 1870s.

CHAPTER 10

The State of Chemistry in Britain in 1846

While there have been many different interpretations of the causes and consequences of the Industrial Revolution, no one has denied that Great Britain led the way in industrialization. Most historians would also agree that France and Germany and other European nations showed few firm signs of catching up with Britain until the 1860s. Indeed, the Paris Exposition of 1867 is often viewed as a crucial marker for Britain's decline from supremacy. Was this how contemporaries saw the European industrial picture? The purpose of this story is to draw attention to a contrary perspective for one area of industrialization, namely the chemical industry, by a German visitor to Britain. Written for a German newspaper in 1846, exactly five years before the 1851 Great Exhibition was mounted to demonstrate the magnitude of Britain's industrial and cultural supremacy, Ernst Dieffenbach's critical remarks offer a valuable cautionary qualification to standard views. For although Dieffenbach confirms what historians have always believed about the significance of the establishment of the Royal College of Chemistry in 1845 in redressing Britain's backwardness in organic chemistry, he challenges our assumption that Britain was well ahead of European nations industrially. Dieffenbach's negative views, here translated, provide a different perspective on the quality of British manufactured chemicals

The Case of the Poisonous Socks: Tales from Chemistry
By William H. Brock
© William H. Brock, 2011
Published by the Royal Society of Chemistry, www.rsc.org

(sulfuric acid, alkalis, soaps. *etc.*) compared with those made in Germany and France in the mid-1840s. Even if this criticism may have been what his German readers wanted to hear, the remarks are still interesting and worth drawing to the attention of historians of chemistry and economic historians.

The name of Ernst Dieffenbach (1811–1855) is unfamiliar except among those interested in the early exploration of New Zealand for whom his *Travels in New Zealand* (1843) is a *locus classicus*. This is, however, the same Dieffenbach whose name appears as a contributor to Justus von Liebig's brave attempt to continue Berzelius's useful annual abstract of the progress of chemistry and allied sciences, for Dieffenbach was a pupil, disciple and friend of Liebig's (Figure 10.1).

Johann Karl Ernst Dieffenbach was born in Giessen on 27 January 1811, the son of a Lutheran clergyman. He matriculated into the Faculty of Medicine at the University of Giessen in 1828,

Figure 10.1   Ernst Dieffenbach (1811–1855), German pupil of Liebig's who helped survey and explore New Zealand between 1839 and 1842. In this dramatic lithograph by the artist Joseph Jenner Merrett (1816–54), Dieffenbach (second left wearing a cloak and hat) witnesses the Maori chief Te Waru denounce his daughter for the murder of a slave. First published in Dieffenbach's *Travels in New Zealand* (1843), this is the only known portrait of the chemist and geologist. (Alexander Turnbull Library, Wellington, New Zealand)

and attended Liebig's lectures and laboratory classes. This was a period of radical student politics and most student fraternities aimed to follow the French in obtaining "unity, justice and freedom". Surviving police records show that Dieffenbach was one of the ringleaders in the storming of the Frankfurt Barracks in April 1833. Subsequently, when the Hessian authorities meted out dire punishments (including hanging), Dieffenbach fled to Switzerland, where he completed his medical training at the University of Zurich in 1835 only to be expelled from Switzerland for subversive activities in April 1836. Although many political refugees were officially pardoned in 1840, Dieffenbach did not believe it safe to return to Germany to see his parents until June 1843. However, it was not until the 1848 Revolution, and again with Liebig's assistance, that he was allowed to resettle in Giessen where, surprisingly in view of his political history, he co-edited the newspaper, *Freie Hessische Zeitung*, with two other of Liebig's former radical students, Moritz Carrière and Carl Vogt. In March 1849 he habilitated at the University of Giessen with a thesis, *Die Aufgaben des geologischen Studium* [The Functions of Geological Studies], and in the following year he was appointed to a chair in mineralogy and geology. Sadly, he caught typhoid fever and died on 1 October 1855, his political activities having severely restricted his scientific output.

While a refugee and after some months in France, Dieffenbach had settled in London, eking out a precarious living from medical journalism and translations of English books into German. It is probable that he had some connection with the medical school at Guy's Hospital, for he certainly became friendly with its Quaker curator and pathologist, Thomas Hodgkin (1798–1866), whose *Lectures on Morbid Anatomy of Serous and Mucous Membranes* (1836) he translated into German. Hodgkin introduced Dieffenbach to Richard Owen, who was at this time Hunterian professor at the Royal College of Surgeons and already at the centre of a circle of scientific patronage. In April 1839, through the influence of Hodgkin and Owen, Dieffenbach was appointed surgeon and official naturalist to *HMS Tory* on a voyage sponsored by the New Zealand Company to survey and explore the islands with a view to their settlement by British emigrants. The ship left Gravesend on 3 May 1839, arriving at Queen Charlotte Sound, the strait between the north and south islands of New Zealand, on 16 August 1839. Following an exhausting survey, during which Dieffenbach became the first European to ascend Mount Egmont (2,518 m), the North Island's highest mountain, he returned to London in January 1842. As they did for Darwin on his return to London in 1836 following the *Beagle*'s circumnavigation, the London scientific community feted Dieffenbach. He was particularly drawn into contact with Charles Lyell and other fellows of the Geological Society.

Stimulated by his previous conversations with Hodgkin, in New Zealand Dieffenbach had become very interested in the Maoris and more generally in ethnological issues. Following his resignation from Guy's Hospital in 1837, Thomas Hodgkin had founded the Aborigines Protection Society in 1836 in order to collect "authentic information concerning the character, habits and wants of uncivilized tribes" and to influence public opinion in such a way that native peoples would not be exploited or civilized. In 1843, Hodgkin, Dieffenbach and others founded the Ethnological Society of London to give focus to scientific concerns rather than the humanitarian ones that were the central motive of the other society. Dieffenbach acted as its first President. Meanwhile, inspired by the success of Darwin's account of the voyage of the *Beagle* (which he was to translate into German at Liebig's suggestion), Dieffenbach wrote up his account of the New Zealand voyage. Later, he was engaged by Liebig's publisher, Eduard Vieweg, first to translate Lyell's *A Second Visit to the United States of North America* (1849), which made critical comparisons between American and British scientific education, and then to translate De la Beche's popular *Geological Observer* (1851).

In 1836, as part of his efforts to build a career and to support himself by journalism and translation, Dieffenbach had contacted Johann Georg Cotta, the owner of a liberal publishing house at Marbach. He had offered to translate the anatomical work of John Hunter which had recently been published in a collected edition by Richard Owen. Although Cotta had rejected the offer at this time, the overture led to Dieffenbach's contributions following his return from New Zealand to Cotta's two newspapers, *Die Augsburger Allgemeine Zeitung* and *Das Ausland*. It was Dieffenbach who encouraged Liebig to republish the "letters" on chemical subjects that had appeared in Cotta's newspaper between 1841 and 1843, and who translated them with the aid of John Gardner (1804–1880), the London apothecary and friend of Liebig's, who was planning to open a College of Chemistry there with the druggist

John Lloyd Bullock (1812–1905). Dieffenbach received £50 from
the German publisher for being the midwife of Liebig's famous
Familiar Letters on Chemistry (1843).

Although there is no evidence that Dieffenbach was ever
involved in the creation of the Royal College of Chemistry, his
friendship with Gardner and Bullock suggests that he was aware of
the plans. Indeed, he appears to have known Sir James Clark, the
Court Physician to Queen Victoria and Prince Albert who was the
driving force in the eventual successful establishment of the College. Soon after Liebig's hurried business trip to Britain in April
1845 in order to further his patents on artificial fertilizers and
amorphous quinine, he wrote to Liebig:

> "I nearly forgot hardly had you left London, when Sir James
> Clark sent a message saying that Prince Albert was desirous of
> making your acquaintance. How much I regretted that you
> had gone! Sir James also had the intention of arranging a
> fashionable ladies' party in your honour. Chemistry has found
> a real patron in Sir James, and it is up to you to establish a
> Chemical Institute in London, if you will let Sir James or the
> prince have your views. I need not emphasize that Clark is a
> real gentleman in every respect."

Liebig was, of course, well aware of the plans for the College,
having been consulted on the matter by William Buckland and Sir
Robert Peel in October 1844. At that time Liebig suggested that the
best place for such a college was not London, but in association
with a provincial agricultural college. Dieffenbach's letter also
complained that he and Friedrich Knapp (who had been sent from
Giessen to James Muspratt's alkali factory in Liverpool specifically
to oversee the production of Liebig's fertilizers) found English
"intellectual life, with the exception of trade and manufacture, at a
low ebb".

Die Augsburger Allgemeine Zeitung of 31 May and 25 June 1846
contained two polemical reports from Dieffenbach, who was
identified as "our man in London". The first, which dealt with
"municipal affairs and agriculture", will only be discussed briefly,
though it is not without interest to historians. Dieffenbach had
been acting as Liebig's principal English agent in persuading
British farmers to adopt his master's mineral manures then being

manufactured under patent by James Muspratt and his sons James Sheridan Muspratt and Richard Muspratt. Dieffenbach's few surviving letters to Liebig show him testing and appraising the respective merits of artificial and natural manures, badgering farmers to order Liebig's fertilizers, finding relations with the Muspratts difficult, and being thoroughly contemptuous of the British business mentality and class that the Muspratts and the capitalist of the venture, the landowner Sir John Wolseley, represented. Dieffenbach probably helped Liebig to write *An Address to the Agriculturists of Great Britain Explaining the Principles and use of Artificial Manures* (1845) which launched the disastrous campaign to sell the fertilizers. The fruitful potentiality of artificial fertilizers is very much the burden of the first article in *Allgemeine Zeitung* which also, however, adumbrated Liebig's interest in the recycling of human sewage by piping it onto fields as fertilizer rather than out to sea. Like Edwin Chadwick, Dieffenbach associated animal and human waste and the contents of midden heaps with epidemic diseases and recommended recycling and proper piped sanitation as weapons in the eradication of diseases such as cholera and typhoid, as well as the reduction of funeral expenses!

Dieffenbach's second brief news report, *Grossbritannien: Chemisches Laboratorium in London* [Great Britain: A Chemical Laboratory in London], was clearly influenced by the highly contentious article on the state of chemistry teaching in Prussia that Liebig had published in *Annalen der Chemie* in 1840. That Dieffenbach had reread this in 1843 is clear from a letter he wrote to Liebig at that time:

"I read once more and very attentively your article on the position of chemistry in Prussia. The situation you touch upon is especially valid for Germany, but less so for England, where the whole trend of the physical sciences has always been to the concrete and strictly inductive. Bacon's philosophy has become second nature to them. What you say about the crisis of faith and its negation brought about by the *Naturphilosophen* would be understood over there only incompletely or not at all."

"A Chemical Laboratory in London" is a stimulating critique of what Dieffenbach saw as the parlous state of chemical teaching in Britain and the ignorance of chemical manufacturers. The situation had at last been remedied, he argued, by the establishment of the Royal College of Chemistry, which had opened in temporary premises in October 1845. But its establishment owed little or nothing to British chemists; rather, as he chauvinistically remarks, it had come about through the achievements of Liebig. Although he does not say so, effectively London now had a "Giessen on the Thames".

The article contains several points of interest and surprise for historians of chemistry. He confirms the view of historians that chemistry at Oxford, Cambridge and Dublin was moribund in the early 1840s. Chemistry at Cambridge would revive only after 1851 when the Natural Sciences Tripos was introduced. At Oxford serious teaching and research did not begin until Liebig's pupil Benjamin Collins Brodie Jr (1817–1880) succeeded Charles Daubeny in 1857. And despite the presence of Liebig's Irish pupil, Robert John Kane (1809–1890), in Dublin, there was little teaching at the Dublin Apothecaries' Hall, the Royal Dublin Society, or at Trinity College.

Dieffenbach's outsider's view of the identities of Britain's greatest chemists—Priestley, Cavendish, Black, Dalton, Davy, Wollaston Brewster and "the incomparable Faraday"—is scarcely surprising. Sir David Brewster may seem an odd candidate among these chemists, but as Dieffenbach's critical reference to these chemists as being devoted to "the physical side of chemical knowledge" suggests, he took a conservative view that the phenomena of light and colour that Brewster brilliantly investigated were part of chemical philosophy.

Overlooking William Hyde Wollaston's claims to innovation with microanalysis, Dieffenbach perceived British chemists as unoriginal when it came to analysis—after all, most new elements had been discovered by Berzelius in Sweden. His reference to William Phillips suggests that the reputation of this overlooked mineralogist deserves reassessment, while he confirms the historian's impression that Thomas Thomson's reputation never recovered from Berzelius's savage critique of his analytical competence in 1829.

Dieffenbach's opinion of British geologists is the same that Liebig had revealed in a private letter to Faraday in 1844 (perhaps Dieffenbach and Liebig had spoken about the matter when they met in 1845), namely that British geologists lacked chemical insights. Above all, Dieffenbach confirms that the British lagged behind in organic chemistry until stimulated by the writings of

Liebig over agricultural and physiological chemistry at the beginning of the 1840s. Pedagogically, he asserts a Pestalozzian philosophy that students need to do practical work because this trains the hand and eye.

It may come as a surprise to read Dieffenbach's judgement that English chemical manufacturers were perceived as backward as early as 1846. We do not know whether chemical manufacturers were, in his view, different from other British industrialists. Whether Dieffenbach was justified in asserting that the quality of alkali was higher on the continent than that manufactured on Merseyside warrants further investigation by historians of chemical industry. What is certain is that this kind of rhetoric, coupled with Dieffenbach's almost throwaway remark at the end of the article that British chemists had largely opposed the establishment of the Royal College of Chemistry points towards the reasons why the College got off to an uncertain beginning. Rhetoric such as Dieffenbach was using, deploring the condition of chemistry in the United Kingdom of Great Britain and Ireland, and the quality of its chemical manufacturers, was undoubtedly counterproductive— just as, earlier in 1830, Charles Babbage's complaints that British science was in decline, had been. Indeed, partly because of Thomas Wakley's strong support for the College in *The Lancet,* both Jacob Bell, the editor of the *Pharmaceutical Journal,* and the editor of the rival *Pharmaceutical Times* were highly critical. It was only following Gardner's dismissal as College secretary at the end of August 1846, and Hofmann's demonstrable brilliance as the College's leader, that the attitude of both pharmaceutical journals changed. In an editorial, Bell noted that the fatal error of the original College Council had been to decry "in unmeasured terms British chemists in general (always excepting a few who had been students at Giessen), and by this discourtesy caused the project to be viewed with distrust or indifference by the parties thus reviled".

Dieffenbach's brief piece of journalism concerning the establishment of what was to become a major chemical institution in London provides an intriguing outsider's view of the progress of British chemistry and chemical industry in the mid-1840s. Even allowing for its chauvinism and polemical exaggeration, it demonstrates once again the powerful position that Liebig had cultivated and established in Great Britain for himself and his pupils. The translation now follows (Figure 10.2).

Figure 10.2 The Royal College of Chemistry whose foundations were laid by Prince Albert in June 1846. The principal teaching laboratory was on the first floor. (Wellcome Images)

TRANSLATION OF "GREAT BRITAIN: A CHEMICAL LABORATORY IN LONDON"

London, 16 June [1846]. I have just returned from the foundation stone laying ceremony by Prince Albert for the new chemistry laboratory in Hanover Square before a circle of handsome ladies and numerous company. Lord Clarendon opened the proceedings with a long, well-made speech. Then the Prince, who has supported this subscription-based chemical institution, and who has through his words and deeds made sure from the beginning that it would obtain public support, completed the old and symbolic act with great dignity. He won the undoubted approval of everyone present for his modest demeanour and his apposite remarks. Samuel Wilberforce the youthful Bishop of Oxford, who yesterday in the House of Lords preached such admirable political morals to the old gentlemen, gave the concluding speech. A temporary laboratory of Dr Hofmann's, a countryman and former assistant of Liebig's, has already been in operation since last autumn, but it attracted so many pupils right from the beginning that a much larger building became an urgent necessity. The foundation of a chemical establishment in London was a thoroughly timely undertaking; it promises to become of great importance for Britain in the future. We Germans will be gratified by the fact that, in order to fill this teaching position, it was necessary to have recourse to German science. For, strange as it may seem, Britain has never had an institution of the kind where the many people for whom a knowledge of chemistry is imperative as a source of their commercial prosperity can obtain practical skills in this exceptionally important science. At the English universities of Oxford and Cambridge laboratories exist in name only; in London, it has only been a short time since regular courses of practical chemical analysis were started at University College and King's College; it is slightly better in the Scottish universities; while nobody is using the fine spacious laboratory in Dublin. It is, of course, true that Britain, despite this neglect, has produced a great many of the greatest names in the history of chemistry: Priestley, Cavendish, Black, Dalton, Sir Humphry Davy, Wollaston, Brewster, *etc.* and in our own day the incomparable Faraday, who is equally famous as an original investigator of the field of magnetism and electricity, as he is for the indefatigable clarity of his lecturing to the young and old, and as the consummate experimentalist of our time. However, for the most part these men have only been concerned with the physical side of chemical knowledge and the British have produced almost nothing distinguished in fundamental analysis. Among their geologists one finds either no, or only ordinary, chemical knowledge. The only British mineralogists are the now deceased William Phillips and the present Professor of Chemistry at Glasgow, Thomas Thomson, of whom the former was principally a crystallographer, while the mineral analyses of the latter (if we are to believe old father Berzelius) are unreliable. In the analyses of inorganic substances, and even more in organic chemistry, Germany and France have jumped ahead of Britain. The new stimulus there in this important branch of chemistry, evident now for some time both in the scientific world as well as in that of the practical agriculturist, is solely and uniquely attributable to Liebig's influence. He aroused general interest in this important branch of chemical science through his reports on the state of plant- and animal-chemistry, and especially through the enthusiastic devotion of his pupils who have meanwhile spread over the whole of Britain. The new institution was, in fact, created for the introduction of Liebig's systematic instruction in inorganic and organic analysis, and the method of teaching is the same as that followed in the master's laboratory at Giessen. Only when a student progresses in analysis from the simplest to the most complex, and obtains a perfect knowledge of chemical phenomena through his own hard work and strenuous efforts with the aid of his own senses, will he finally gain sufficient assurance to make exact observations in the immense field of science, or bring about improvements or inventions in one of the very many opportunities presented to the chemist in industrial Britain.

At the end of the day, genius finds its own way forward, without the aid of school or method; but what Britain needs as a matter of priority is the general knowledge of chemistry as the first and most preferable subject of study for its manufacturers. Experts in this field know very well how far behind the continent Britain is in many of its expanding trades, in the manufacture of soda, in glass and in soap manufacture, in the dyeing and other chemical industries. It is only the extraordinary diversity of her natural resources and her wealth that explains the ease with which she can acquire inventions from abroad and utilize them on a large scale. British manufacturers' world trade and the ability even of chemical producers to compete with all nations is not attributable to any dissemination of knowledge of the principles upon which they proceed. It is well known that the Germans and French produce a better kind of sulfuric acid and soda and maintain a larger percentage yield than British manufacturers. But in their case it is the lack of a spirit of enterprise, their slowness in putting new ideas into action, their suicidal inhibitions and shortage of capital, because of mutual distrust, that hinders growth and expansion of their chemical industries. Citizens of the world that we all are, we wish Britain well and hope that she will in good time kindle her torch on the hearth of German science. This will not come about by an order from above, mind you, but through collaboration between individuals for a common purpose—individuals of whom not a single one will profit directly from an enterprise to which they contribute their time, labour and money. The moral for Germany we keep to ourselves!

It is thanks to Prince Albert, and especially to the indefatigable Queen's Physician, Sir James Clark, that the Chemical Institute has been able, within the space of two years, to overcome all the obstacles put in its way, the chief of which has been the opposition of British chemists who, with the exception of Faraday, have been hostile towards it. [Translated from *Allgemeine Zeitung*, 25 June 1846, p. 1404]

CHAPTER 11

The Laboratory Before and After Liebig

The laboratory has been an essential factor in the evolution of modern science. The coming of the laboratory was largely a nineteenth-century phenomenon, for it was in that century that the laboratory became a symbol and indicator of the professionalization of science, as well as a utilitarian tool of teaching and research. Architecturally, because the scientific community wished to assert its status and identity, its professional home had to be impressive—indeed, a palace or cathedral, or what Pasteur called "temples de l'avenir", temples of the future.

The "Chemical Revolution" associated with the brilliant work of Antoine Lavoisier at the end of the eighteenth century is usually thought of in terms of the theoretical changes he introduced. However, these changes of language and explanation of chemical processes were also accompanied by considerable changes in laboratory practice. Hitherto the laboratory, or "elaboratory", had been effectively a small factory where metals were assayed, drugs were compounded, or natural dyestuffs prepared. From the 1750s onwards, particularly in Sweden, analytical methods using the blowpipe or reagents (such as the stinking hydrogen sulfide gas that produced distinctive coloured precipitates with different metallic salts) were beginning to be systematized into a precise methodology that could be used to investigate the composition of unknown

The Case of the Poisonous Socks: Tales from Chemistry
By William H. Brock
© William H. Brock, 2011
Published by the Royal Society of Chemistry, www.rsc.org

inorganic substances. By becoming systematic, qualitative and quantitative analysis became teachable and useful not only in medicine but also in agriculture and industry. Such systematization permitted the miniaturization of the laboratory in the form of portable chests that could be carried to the bedside by doctors or into the field by mineralogists. Several entrepreneurial teachers, such as David Boswell Reid in Edinburgh and London, and Amos Eaton in Massachusetts, also used systematic spot tests made possible by portable laboratories to instill analytical skills in children and young adults by rote teaching methods. The child's chemistry set and the "test tube" made ubiquitous in chemistry through Michael Faraday's laboratory handbook, *Chemical Manipulation* (1827), are enduring memorials to this movement.

In the long run, however, and mainly because of the development of organic analysis, it was the expansion, rather than the miniaturization, of the laboratory that was to be the formula for organizing scientific work. The development of new laboratory skills in connection with organic analysis (and later synthesis) brought about the emergence of a distinctive chemical profession. It was primarily in Germany, rather than in France or Britain, that this laboratory revolution of chemical practice took place through the creation of a large number of private pharmaceutical–chemical schools that aimed to train pharmacists to recognize the adulteration of food and drugs by the application of the new science of chemical composition. Liebig's first laboratory at the University of Giessen in 1825 was to develop from such a private school.

By the 1860s the laboratory was no longer a cramped workshop that resembled a blacksmith's or a domestic kitchen. Instead it had become a specially constructed room or rooms filled with benches and tables at which a number of people could simultaneously carry out analyses and small-scale experiments. This revolutionary change was well noted by the French chemist, Adolphe Wurtz, writing in 1869:

"On entering a [German] laboratory, you no longer see those crude contraptions, that bizarre equipment, that smoke-blackened hearth, those furnaces vomiting cinders and reddening the features of the operator. In a spacious room, with ample light and air, you will find large tables facing windows so as to receive the daylight directly."

By then the experimental method of chemistry teaching and investigation was also providing a model for other sciences, beginning with physics and physiology. While this change cannot be attributed entirely to one man, it was Justus von Liebig (1803–1873) who was, above all, associated with this transformation of the laboratory. Germany was to remain dominant in laboratory design. For example, American laboratories continued to rely on German designs, equipment and chemicals until the First World War. In any survey, therefore, it is convenient to look at the laboratory before and after Liebig.

LIEBIG'S CONTRIBUTION

So far, we have established only that the early laboratories (many of which continued in use throughout the nineteenth century) were small in scale. If they were used solely to prepare lecture demonstrations or for the personal research of one person (as in Faraday's case), or for routine commercial analyses of water, ores and soils, they were inconvenient but adequate. But as soon as practical study became recognized as part of teaching, either for reasons connected with liberal education or for pure utility, expansion of laboratory space became essential. Expansion posed problems, as well as the opportunity to rethink the design of laboratories.

Although teaching laboratories had existed in some German universities before Liebig, it was the tiny University of Giessen that established a model for laboratory design for the next fifty years. Here Liebig's perfection of combustion analysis of organic compounds in 1831 and his ability to develop systematic ways of training students and colleagues to pursue research questions that he laid out (and his subsequent international fame) became a model for other nations to copy. Moreover, by demanding and obtaining a subsidy from the government to cover the expenses of running a laboratory, Liebig overcame the assumption that laboratory instruction was a professor's personal expense. Similarly, as a successful model of what a laboratory ought to be like, it and its successors acted as a template for the earliest physics and biology laboratories in the 1870s. Liebig's description of his laboratory in 1842 also gave rise to an architectural literature, with proud descriptions appearing of magnificent new laboratories at Oslo (1854), Heidelberg (1858), Munich (1859), Marburg (1865), Bonn and Berlin (1866), Leipzig (1868), Budapest (1872), Aachen (1879), Graz (1880) and Manchester (1906).

From 1825 until its first enlargement in 1835, Liebig's laboratory was the guardroom of a disused barracks on the town's boundary (now the Ludwigstrasse) close to the railway station that was to arrive in the 1850s. Apart from a spectacular open colonnade in Greek style, where dangerous reactions could be performed in the open air, Liebig had an internal laboratory space of little more than 38 square metres. This was divided into a small room that served for lectures, an unheated broom cupboard that served as a balance room and store, while the remaining space was filled with furnaces fanned by bellows and work tables for both Liebig himself and the eight or nine students who might work there at any one time. On the floor above, in equally cramped conditions, lived Liebig, his wife and their five children. Apparently the family's laundry was done in the laboratory in the rare moments it was empty.

By 1833, following some hard bargaining with the Hessian government, Liebig amalgamated his private pharmacy school with his official university chemistry course, by which time he was teaching ten to fifteen pharmacy students and three to five chemistry students per year. Over the next two decades, however, the number of students reading chemistry was to exceed those in pharmacy. By 1852, when he left Giessen for a quieter life at the University of Munich, more than 700 students of chemistry and pharmacy had passed through his hands.

Such numbers could only be dealt with in a much larger laboratory than the premises Liebig had been given in the 1820s. Leaving aside a small extension built in 1835, it was not until 1839 that he and the Hessian state architect Paul Hofmann (whose son, Wilhelm, was soon to become Liebig's favourite German pupil and another great laboratory builder) obtained funds from the government to enlarge his facilities to include a new lecture theatre and two large laboratories for pharmacy and chemistry students. This new chemistry laboratory was fitted with glass-fronted cupboards from which the fumes from dangerous reactions and evaporations could be led directly into the outside air through a special chimney. Such "fume cupboards", which had hitherto been found only in the private laboratories of Gay-Lussac and Thenard in Paris (where Liebig must have first seen them) and the pharmacist W. H. Pepys in London, became standard laboratory furniture all over the

Figure 11.1   Part of Liebig's teaching laboratory at the University of Giessen in 1842. From a drawing by Wilhelm Trautschöld. The chemist on the right wearing a hat is Hofmann. (Wellcome Images)

world. Connected to a chimney and arranged with a glass shutter at the front to admit a current of air, dangerous operations could be safely confined to the cupboard without poisonous fumes entering the laboratory. They must have saved countless chemists' lives (Figure 11.1).

Liebig now had the best laboratory complex in the world: a lecture theatre for 70 people, an analytical laboratory for 20, another pharmaceutical training laboratory for 20, a library, his own private research laboratory, as well as other specialized facilities such as a balance room. The famous lithograph of the analytical laboratory made by Wilhelm Trautschold and Hugo von Ringen in 1842 is recognizably a modern laboratory. We note familiar benches with cupboards stretching along the walls by the windows and shelves for reagents, while under the benches were recesses for containers from which water could be drawn through a tube. (Running water from pressurized mains was to come later.) A large stove stood in the middle of the room to heat the laboratories in the winter. Against another wall there was a large drying kiln containing sliding windows so that items could be dried while protected from dust and contamination. Pipes and flues from this stove also led into other rooms for heating and for conducting similar drying procedures. Four tables (later benches) were arranged in aisles in the central space of the laboratory, and fume cupboards at one end. A "Familius" or caretaker is also illustrated—a reminder that a successful laboratory would in the future need the assistance of a technician or laboratory assistant.

AFTER LIEBIG

Liebig's method of teaching by practical exercises and direct research was tremendously influential. His methods and laboratory designs, as well as those of his pupil Hofmann, were copied and emulated by other state universities and polytechnics in Germany, Austria, Switzerland and Hungary. These in turn influenced the wave of laboratory building in France, Britain and America in the 1870s. As Kolbe noted:

> "Whoever undertakes to build a chemistry laboratory is taking on no small task since, obviously, one has to achieve something as good as the existing institutions, while avoiding all their mistakes."

The principal problem presented by chemical work is fairly obvious. Chemists produce obnoxious and sometimes dangerous fumes (British school children have nicknamed them "stinks" ever since the 1870s.) Laboratory atmospheres need ventilation beyond that supplied by open windows. The latter might suffice for one experimenter, but not for a gathering of twenty to thirty students. Prior to the 1860s, the usual way of dealing with these ventilation problems was either (as Liebig did) to use window ventilation backed up by chimney hoods, or to put a laboratory in the basement adjacent to the main drains and chimney (as at the Royal Institution and, later, at the Pharmaceutical Society). The Pharmaceutical Society's laboratory (1845) was actually built in the back garden of a house in Bloomsbury Square, with skylights, and sliding vents leading to the main chimney where evaporations could be conducted. Drains from the sinks led directly into the town sewers. Hydrogen sulfide gas was also on tap and in its own ventilation shaft—a plan adopted by Henry Roscoe and Henry Armstrong when designing their respective laboratories at Manchester and the Central Institution in London in the 1860s (Figure 11.2).

The Pharmaceutical Society's design worked well enough for the chemists in it, but drifting smells made life uncomfortable for those

Figure 11.2   The Pharmaceutical Society's basement laboratory in Bloomsbury Square in 1845. Note the individualized work spaces and wall traps for wastes. Because of smells and drainage problems, the laboratory was later rebuilt on the top floor. (Science & Society Picture Library)

in the rooms above. Consequently, in the 1880s, the laboratory was moved to the top storey of the building. In buildings where there were laboratories for several scientific disciplines this became the preferred solution: put the chemists next to the sky! Even so, by the 1870s, it was clear that more powerful ventilation systems were necessary and using electric fans on the roof that sucked air through the ventilation shafts solved this problem. Ventilation hoods, sand baths, drying ovens and water supplies for filter pumps became essential design specifications (Figure 11.3).

Particularly important were the designs that Hermann Kolbe made in 1868 for chemistry laboratories at the University of Leipzig serving 130 students. He was influenced by the design of the Birkbeck Laboratory which opened at University College London in 1846. Kolbe's innovation was the furnishing of the laboratories with double-sided benches for four workers with bottle racks and a sink with water taps that could either be placed at the side or incorporated into table top of the bench itself, together with gas

THE NEW LABORATORY, AT UNIVERSITY COLLEGE.

Figure 11.3   The Birkbeck Laboratory at University College London opened in 1846 to commemorate George Birkbeck's services to scientific education. George Fownes, the first professor of practical chemistry to run the laboratory, was succeeded by Alexander Williamson in 1849. The laboratory became a botanical and pathological laboratory in the 1890s. (Science & Society Picture Library)

and steam supplies. The double bench dimensions of 5 m × 1.5 m × 0.9 m with a 1.2 m separation between them became and remained a standard until the 1960s. Kolbe's raked lecture theatre (copied from the grand auditorium Liebig built for himself at the University of Munich in 1853 and by those Hofmann designed for Bonn and Berlin in 1865) also became *de rigueur* in other new institutes, and replaced the uncomfortable and ill-lit lecture rooms of the past (Figure 11.4).

By the 1870s improvements in mining ventilation technology and household sanitation made it possible to enlarge laboratories safely and efficiently. There was a transformation of chemical and physical techniques made possible by the introduction of gas lighting and heating, water pumps and condensers. The rise in importance of organic chemistry entailed more delicate control of temperatures

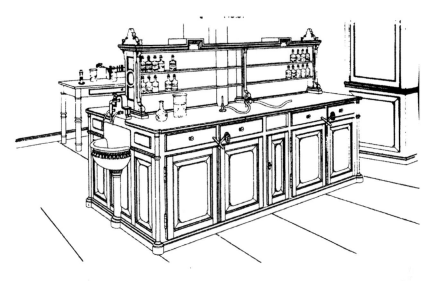

Figure 11.4 Design for a four-place island chemistry bench by Hermann Kolbe at the University of Leipzig in 1868. This design, with sinks at either end and integral gas supplies and shelved reagents, became adopted all over the world. From E. C. Robins, *Technical School and College Building* (1887). (Brock)

and, often, the production of deadlier fumes. The emergence of new physical instruments such as the spectroscope, electric dynamos and high-pressure filtration systems, distillation under reduced pressure, and the vogue for exact measurements and standardization of units, all became the epitome of modern experimental styles.

There was no gas supply in Liebig's laboratory and analytical combustions were still done by means of charcoal. Gas supplies were first introduced in Robert Bunsen's laboratory at Heidelberg in the 1850s. Coincidentally, Bunsen's contemporary work on the colour of flames led him to design the Bunsen burner, which quickly replaced the charcoal furnace for most heating purposes. Gas furnaces for combustion analysis also appeared in the 1860s and electric combustion furnaces with platinum heating elements were available by the 1890s. Like electricity later in the century, gas lighting also extended the scientists' working day by making research in the evenings possible. When Hofmann came to design the prestigious cathedral-like chemistry laboratories at Bonn and Berlin in the 1860s, he introduced gas heating, and Kolbe and his successors everywhere followed this. Distillation techniques were also transformed from simple retort and receiver into the flask with attached Liebig condenser (developed by pharmacists but popularized by Liebig). Rubber tubing for water feeds and the use of distilled water for washing precipitates, and devices (like Kipp's apparatus) for the constant supply of hydrogen sulfide gas, were introduced everywhere (Figure 11.5).

The rise of physical chemistry at the end of the nineteenth century brought further changes. New instruments and techniques such as pH meters, ultra-violet and infra-red spectroscopy, spectrophotometers, X-ray cameras, mass spectrometers and chromatography began to offer new, simpler and faster methods of analysing materials and directly assigning a structure to them. Such physical methods had completely replaced wet and dry methods of analysis by the 1960s, so that today the traditional laboratory bench of Liebig's and Kolbe's time has lost its scaffold of reagent shelves and returned to a table-like appearance.

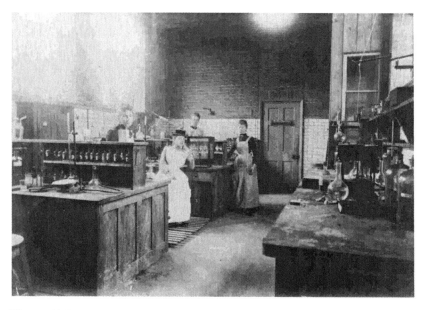

Figure 11.5   Ida Freund's women's laboratory at Newnham College, Cambridge, in the 1890s. Sinks are now incorporated within the 'Leipzig' bench. Freund (see chap. 26) is the figure wearing a hat. (Newnham College)

CHEMISTRY'S INFLUENCE ON OTHER SCIENCE LABORATORIES

The discipline of physics separated completely from chemistry only after 1850. Until the concepts of energy and electromagnetism were fully established, heat, light and electricity were perceived as areas of experimental chemistry. That is why figures who had been trained in chemistry, such as Faraday, Regnault or Andrews, seem more like physicists to us today. By the 1870s, several chemists who had been trained in the Liebigian tradition of organic chemistry began to exploit their laboratory skills to create teaching programmes in the new physics. A powerful incentive here was the growing market for engineers in electric telegraphy. Such engineers needed training with electrical apparatus and electrical experiments. One such pioneering and influential laboratory was that designed by William Ayrton and John Perry for the Japanese government at Tokyo in 1877. However, the equipment for physical experiments involving delicate optical and electrical instruments was considerably more expensive than that used in teaching elementary chemistry. The solution, developed and widely publicized by Edward Charles Pickering (1846–1919) of the Massachusetts Institute of Technology in Boston, was to devise a rota of physics experiments, apparatus and instruction sheets for students to work in pairs. Students would then proceed in turn from one table to another, where a different set of apparatus was to be found, over a period of months. Tables, rather than benches, fitted with water, gas and electric sockets, brackets and ceiling hooks for experiments with pendula and springs, and blinds to exclude light, therefore became the hallmark of the adaptation of the chemistry laboratory for physics.

For more advanced work, physicists demanded elaborate (and hence costly) building foundations and specialized piering to enable precise vibration-free measurements with balances and delicate galvanometers, or rooms completely free of iron for magnetic research, or even a tower for experiments on pendula and free-falling bodies! Mechanical engineers, too, required rooms adequate to house very heavy specialized machinery.

In the biological sciences, microscopy had become a tool for identifying chemical substances by the 1850s. Hofmann's new laboratories at Bonn and Berlin had special windowed areas for microscopy. Apart from an animal house, a physiology laboratory dedicated to dissection and vivisection was easily adapted from a workroom modelled on a chemistry or physics laboratory. The rise of bacteriology in the hands of Louis Pasteur and Robert Koch stimulated the establishment of health-related laboratories. Pasteur's laboratory at the Ecole Normale Superieure, Paris, was recognizably a converted chemistry laboratory still dominated by furnaces and hoods, though the benches were used for culture bottles rather than chemicals.

By the 1870s the chemistry laboratory had spawned laboratories for training and research in physics, engineering, physiology and bacteriology. Each discipline presented its own particular design problems: ventillation and drainage in chemistry; vibration in physics; strong foundations in engineering; animal housing in physiology; and air purity in bacteriology. Consequently, laboratories demanded highly specialized architectural planning. Such architects were, however, only too happy to embellish their commissions. In the "Cathedrals of Science" they designed they were able to provide a statement concerning the role and function of science in modern industrial societies.

CHAPTER 12

The Chemical Origins of Practical Physics

Apart from art and music, what distinguishes science, medicine and engineering from all other forms of vocational training is the laboratory/workshop location in which it takes place. This essay looks at the relation between nineteenth-century physics and chemistry and considers the development of practical physics teaching in the UK and USA.

There is a remark attributed to Robert Bunsen by his English pupil Sir Henry Roscoe: "Ein Chemiker der kein Physiker is, ist gar nichts"—a chemist who is not a physicist is nothing. Although a modern physicist might well construe this as confirmation that chemistry is merely a part of physics, chemists will accept it in the spirit that Bunsen intended, namely that chemists should also be natural philosophers with a broader perspective and agenda than that of mere analysis: chemists explain physical as well as chemical properties and to do so they must needs grasp and use the principles of mathematics and physics.

In fact, however, as many scholars have pointed out during the last decade or so, the discipline of physics as we know it today only emerged quite late in the nineteenth century, as reflected by the fact that both British and French Physical Societies were only founded in the 1870s. Before then, physics (or strictly speaking, practical physics as opposed to mathematical physics which did have its own

The Case of the Poisonous Socks: Tales from Chemistry
By William H. Brock
© William H. Brock, 2011
Published by the Royal Society of Chemistry, www.rsc.org

autonomy) was very much a part of chemistry, as early nineteenth-century textbooks confirm. As late as 1867, the exciting contents of John Cargill Brough's short-lived weekly journal *Laboratory* referred solely to chemical and pharmaceutical spaces, not physical ones.

Put another way, the imponderable bodies of heat, light and electricity remained part of chemistry until subsumed within the concepts of energy and electromagnetism in the 1850s and 1860s. It is not surprising, therefore, that many nineteenth-century chemists who implicitly followed Bunsen's later aphorism in their research activities seem more like physicists than chemists, when seen from our later disciplinary perspective of physics. Faraday is the most obvious case, and others like André-Marie Ampère, Henri Regnault, Gustav Magnus and Thomas Andrews readily spring to mind. Less well-known are those cases of chemists who exploited their practical laboratory skills to create teaching programmes in physics at the very time when the subject was emerging as an autonomous discipline in the 1870s. As George Carey Foster reminisced in 1914: "the transformation ... from chemist into physicist, was a fairly common phenomenon". This essay is about two of these chemists turned physicists.

Like the 1960s and 1970s, the 1860s and 1870s were a golden age for new laboratories and academic buildings in Great Britain. In 1866 George Carey Foster began to teach practical physics at University College—to be followed three years later by W. G. Adams at rival King's College. Between 1873 and 1874 the Cavendish physical laboratory was opened at the University of Cambridge, with fittings designed by James Clerk Maxwell based upon the experiences of the Clarendon laboratory at Oxford and of William Thomson's laboratory at Glasgow. In 1874 laboratories were opened at the new Royal Naval College at Greenwich, and science colleges were opened at Bristol and Leeds. In 1878 Alfred Waterhouse's gothic University College opened in Liverpool, with its extraordinary tiered chemical laboratory designed by the chemist James Campbell Brown. Sheffield responded with Firth College in 1879 and the decade ended in 1880 with the opening of Mason's College in Birmingham and the beginning of the new City & Guilds of London Institute's ambitious building programme for technical education—completed in 1884 as Finsbury Technical College in the city of London and in 1885, at South Kensington, with Waterhouse's elegant Queen Anne-style Central Institution.

As implied, although the bulk of this building activity was devoted to opening larger and better facilitated chemistry laboratories, the decade also saw the emergence of the physics laboratory for undergraduates and certificated students, as opposed to a laboratory space like Thomson's in Glasgow and Maxwell's at Cambridge in which advanced (postgraduate) students were encouraged to help a professor with research problems. The pioneers here were undoubtedly two former chemists, George Carey Foster and Frederick Guthrie, the one at University College London, the other at the Royal School of Mines (or Normal School of Science as it became) in South Kensington.

But how and why did these two men turn aside from promising careers in organic chemistry for uncertain ones in physics teaching? Let's examine each man in turn and look for any common thread.

GEORGE CAREY FOSTER (1835–1919)

A shy nervous man and son of a Lancashire calico printer, the key to Foster's career was undoubtedly the great theoretical chemist of the 1840s and 1850s, Alexander William Williamson, with whom Foster studied chemistry before briefly acting as his personal assistant from 1855 to 1858. In 1858 Williamson packed Foster off to the continent to study with Kekulé at Ghent, Jules Jamin in Paris (significantly a Professor of Experimental Physics at the Sorbonne who was much interested in electrical problems) and the young Privatdozent, Georg Quincke at Heidelberg—a pupil of Bunsen's who was to spend a lifetime devoted to measuring and collecting data on the properties and constants of materials (Figure 12.1).

From these experiences Foster returned to Britain in 1860 not only *au fait* with Kekulé's structural theory of organic chemistry, but with a deft hand in French and German experimental physics. Between 1857 and 1867 we find Foster addressing the British Association for the Advancement of Science on organic nomenclature (for example, he recommended the term *isologous* to indicate a difference of H_2 in a carbon series, on the analogy of Gerhardt's *homologous* for a difference of CH_2) and on "recent progress in organic chemistry". He also wrote up the research he had done at Ghent with Kekulé on hippuric and piperic acids. On his return Williamson found Foster a position in the private consulting laboratory of Augustus Matthiessen (another pupil of

Figure 12.1 George Carey Foster (1835–1919) trained as a chemist before becoming professor of experimental physics at University College London where he also served as Principal from 1890 to 1904. (Science & Technology Picture Library)

Bunsen's who had emigrated to England in 1857) where Foster produced technically excellent work on the constitution of the alkaloid, narcotine.

In 1862, again due to Williamson and to his Scottish contacts, Foster was appointed to a chair at Anderson's College in Glasgow, not however in chemistry, but in natural philosophy. Perforce, therefore, Foster was forced to lecture on those aspects of physics that he knew best from his training with Williamson as a chemist, and Jamin and Quincke as a physicist, namely heat and electricity. These were both subjects that he contributed articles on to Henry Watts' *Dictionary of Chemistry* in 1863—volumes that were hailed by Crookes in *Chemical News* as "the greatest work which England has ever produced".

While Foster was in Scotland, and building on the back of the introduction of the BSc degree within London University, Williamson and other senior professors at University College London persuaded the decrepit Professor of Natural Philosophy, Benjamin Potter, to take early retirement. The Natural Philosophy chair was then divided into chairs of mathematical physics (awarded to the geometer, Thomas Archer Hirst, who had also studied chemistry with Bunsen at Marburg) and experimental physics, which went to Foster in October 1865. It was at this point, in 1863, that Williamson's chemistry syllabus, which had hitherto contained a good deal of elementary heat, electricity and optics, now became almost exclusively chemical because these subjects were now to be taught separately by Foster. The age of teaching physics separately had begun in Britain as a by-product of London chemists' successful campaign to get science degrees awarded by the University of London. Coincidentally, of course, such a change also released valuable time for the rapidly increasing amount of knowledge of organic chemistry that chemists like Williamson wanted to teach.

We need not follow Foster's spatial difficulties in teaching a physics of quantitative precision. Suffice to say that initially he had to make do with a room adjacent to the lecture theatre before gaining a couple of workshop/laboratory spaces in the basement of University College. Under these cramped conditions, and working for the British Association Committee on electrical standards, in 1872 he transformed the Wheatstone bridge into a precision instrument usually known in Britain as the Carey Foster Bridge. It is important to note that because apparatus was expensive, Foster and his mechanic (William Grant) often designed and made their own apparatus.

In the 1870s Foster campaigned for physics teaching in schools, wrote and edited texts, and in the 1880s he argued strongly for the importance of practical physics and the value of accurate measurement as a foundation for research. His reward came in 1893 when the College erected the Carey Foster Physics Laboratory (destroyed in the Second World War) and the knowledge that he had trained the next generation of physicists such as William Ayrton, Oliver Lodge and Ambrose Fleming. Foster retired in 1898, only to be appointed Principal (Provost) of the College, where he played a leading role in the negotiations that led in 1900 to the reconstitution of the University of London as both a teaching and examining body. One pupil recalled:

"His nervous manner prevented Carey Foster ever becoming a good lecturer, and his failure in this respect was due, in addition, to a conscientiousness that made it difficult for him to be content with a simple statement that he knew to be only an approximate expression of a truth, and at the same time made him reluctant to adopt the customary method of illustrating physical laws by the use of simple, although entirely imaginary, experimental data. In place of these, his illustrations would often consist of the actual results of laboratory measurement, and the younger students, unless they were of a rather exceptional type, were apt to lose both attention and interest in the details of laborious computation."

His patience in showing students how to conduct their experiments often involved him taking over completely. His character was not unkindly, but shrewdly, summed up by Oliver Lodge in the phrase: "He was far from fluent, and was so conscientious about expressing himself correctly that sometimes he failed to express himself at all".

FREDERICK GUTHRIE (1833–1886)

If Guthrie is remembered at all by historians of science and education, it is the terrible portrait of him painted by H. G. Wells in his *Autobiography* (1933) that sticks in the mind. Wells trained to become a secondary school science teacher at the Normal School of Science in South Kensington in 1885. After praising Huxley's biology course for its abiding interest and relevance, Wells writes:

"Now Professor Guthrie, the Professor of Physics, was a man of very different texture from the Dean [Huxley]. He appeared as a dull, slow, distraught, heavily-bearded man with a general effect of never having fully awakened to the universe about him."

Like James Watson's unflattering portrait of Rosalind Franklin in *The Double Helix*, Wells was, I think, trying to give an honest impression of his feelings as a science student forty years earlier; he admits that he only learned later that Guthrie was suffering from throat cancer at the time of these impressions in 1885. But unlike Watson, Wells, while admitting that he was an unruly and unprincipled student, was unforgiving. Guthrie's cancer had only "greatly enhanced the leaden atmosphere of his teaching". "Quite apart from that, he was not an inspiring teacher ... to put it plainly, [Guthrie] maundered amidst ill-marshalled facts. He never said anything that was not to be found in a textbook" (Figure 12.2).

Faced with such a brilliant demolition job (but note that Wells failed his Associateship), the reader may find it difficult to believe that Guthrie, along with his colleagues at South Kensington,

Figure 12.2   Frederick Guthrie (1833–1886) trained as a chemist before becoming professor of physics at the Royal College of Science in South Kensington (now Imperial College). He founded the Physical Society (now the Institute of Physics) in 1873. (Science & Technology Picture Library)

T. H. Huxley and Edward Frankland, and later Henry Armstrong, William Ayrton and John Perry, was one of the great innovators of Victorian science teaching. What are we to make of Guthrie's dedication, as recorded by Lodge: "There was a time when Guthrie lived a curious life; he would not leave his laboratory, even at night. He had a hammock rigged up, and used to live in the laboratory".

Guthrie, who was probably of Scottish ancestry, was the son of a London tradesman. Two years older than Foster, like him he was educated at University College School and the College before studying chemistry with Thomas Graham and Alexander Williamson, as well as Henry Watts, who had been his private tutor until the age of twelve.

Like Foster, Guthrie had gone to Germany in 1854 to work with Bunsen at Heidelberg and then under Hermann Kolbe at Marburg, which he described to his friend Henry Roscoe (whom he had met at University College) as "this dreary valley of desolation". He seems, however, to have got on with Kolbe very well, the latter describing him as a "well-educated man—very industrious and extremely nice". What Guthrie made of Kolbe's unsuccessful use of him to try to refute Williamson's theory of etherification electrolytically would be interesting to know. The Marburg doctorate was earned successfully in 1856 for an experimental study of amyl ether and led to some eight papers on organic chemistry, including one on the first preparation of mustard gas in 1860. In view of Foster's work on narcotine, it is intriguing to note that Guthrie was the first chemist to point out the therapeutic action of amyl nitrite (the nitrous acid ester of isoamyl alcohol), though its use as a vasodilator in heart disease was not proven until the work of the pharmacologist Thomas Lauder Brunton in 1867. Brunton's studies were directly inspired by those of Alexander Crum Brown, the successor to Lyon Playfair at the University of Edinburgh. Guthrie had assisted Playfair in Edinburgh between 1859 and 1861, also serving in a similar capacity with Frankland at Owens College, Manchester, from 1856 to 1859. Everything seemed to point Guthrie towards a successful career in chemistry.

However, in 1861 Guthrie made an extraordinary career move, inspired perhaps by his brother's growing success as a teacher in South Africa. Along with Walter Besant, the future novelist and literary critic, Guthrie became a Professor of Natural Philosophy at the Royal College in Mauritius. The former Ile de France, Mauritius had become a British colony after 1811 as a result of the Napoleonic wars. When Charles Darwin visited the island in May 1836 during his voyage on *HMS Beagle* he found cultivated fields, sugar cane plantations, bookshops, opera houses and tarred roads; this was confirmed by Besant who writes of the island as being "a gay and sociable place" where youth predominated. Besant writes of Guthrie as:

"A man of infinite good qualities. He was my most intimate friend from our first meeting in 1861 to his death in 1886. It is difficult to speak of him in terms adequate. He was a humourist in an odd, indescribable way; he did strange things gravely; he was a delightful donkey in money matters; when he drew his salary—£50 a month—he prepaid his mess expenses, and then stuffed the rest in his pocket and gave it to whoever asked for it, or they took it. Hence he was popular with broken down Englishmen of shady antecedents who hung about Port Louis. He never had any money; never saved any; always muddled it away."

That was probably why he sometimes slept in the laboratory! Besant continues:

"Like many such men, he was not satisfied with his scientific reputation; he wanted to be a poet. He published two [pseudonymous] volumes of poetry, both with the same result. He was also clever as a modeller, but he neglected this gift. He did some good work in the colony in connection with the chemistry of sugar cane."

Guthrie's two poetic attempts are not without interest, for their subject matter on both occasions was the outsider—the Jew and the gypsy. They were both published under the pseudonym of Frederick Cerny; whether the choice of name was significant, we cannot say. *The Jew* (1863), a Miltonesque and Dantesque meditation on the problem of evil, seen through the eyes of a Jew who offered no solace to Christ on the march to Calvary, is not without some fine moments, and may be interpreted as a defence of "seeking knowledge" in order to improve a sinful and intransigent world. A decade later, drawing upon one of the stories in George Borrow's *Gypsies of Spain*, Guthrie published an illustrated two-act metric drama, *Logrono* (1877). This passionate story of a scholar, a gypsy and a lascivious Count, surrounded by a chorus of courtiers, burghers, peasants and flower girls, reads like a libretto for a tragic opera. One cannot help regretting that Bizet, Verdi or Puccini never came across the script.

Besant's reference to Guthrie's carelessness over money is interesting. Like the Anglo-German chemist, August Wilhelm Hofmann, Guthrie married four times (did three wives die in childbirth?) and his widow and her three step-children were left in penury in 1886. Thomas Huxley launched an appeal in *Nature* and Guthrie's widow eventually received a pension from the civil list.

It is Besant who also informs us that, already in Mauritius, Guthrie was a confirmed agnostic "who thought it his duty to learn such of the secrets of nature as he could, and not trouble himself about speculations as to the secrets of life, either before the cradle or after the grave". This does much to explain how Guthrie got on so well with Tyndall, Frankland and Huxley at South Kensington. Guthrie was one of the new science professionals, convinced that science, not religion, was the key to human progress and happiness. Like Tyndall and Huxley, he lectured regularly to working class men.

The Royal College in Mauritius took University of London examinations, but in the absence of a proper chemical laboratory there, Guthrie was thrown upon his own resources, investigating the physics of droplets and bubbles, analysing local river water and commenting upon the quality of sugar cane. As Besant makes clear in his autobiography, the professoriate was unhappy with the Principal, an ex-Austrian army officer who continually upset the local French population. Eventually Besant complained to the island's Governor, who carried out a Commission of Inquiry. Although the Principal was eventually removed, his reinstatement in 1867 brought about Besant's and Guthrie's resignations and return to England.

Back in England, Guthrie landed on his feet teaching science at Clifton College, Bristol, the proprietary school which was also later to have John Perry, Arthur Worthington and William Shenstone on its staff, and Eric Holmyard in the twentieth century. Here he polished off a paper on heat which he sent to John Tyndall, who arranged for its publication and who saw to it that Guthrie succeeded him at the Royal School of Mines in Piccadilly a year later. In 1872 the School was to move to South Kensington, where Huxley, Frankland and Guthrie had improved laboratory facilities for teaching practical biology, chemistry and physics. As a Normal School of Science, Guthrie and his colleagues not only trained would-be science teachers like H. G. Wells, but under the sponsorship of the Department of Science and Art, they were able to set up annual summer schools to improve the practical teaching skills of existing science teachers.

Guthrie joined the Chemical Society in April 1868 and promptly described a voltastat to deliver constant voltage from a battery, and offered an intriguing paper "on graphic formulae". The latter suggested the replacement of Crum Brown's recently introduced graphic formulae by pictorial geometrical symbols that indicated combining power. At the meeting Guthrie's suggestion was treated with derision by William Odling and led to some entertaining correspondence in *Chemical News* and, amusingly, to some buffoonery at the B-Club to which both Foster and Odling belonged. In reply, Guthrie showed his sense of humour in retorting that Odling's own dash valence notation was in reality a system of graphic formulae.

"Dr Odling appeared shocked at the idea of an atom of nitrogen supporting three 'sticks' one in each hand, and one on its head. Strange objection from one who years ago trained his atom of nitrogen the much more difficult acrobatic feat of balancing simultaneously three sticks on the tip of its nose, -N'''."

While on the subject of Guthrie's sense of humour and geniality even when under attack, let me mention two other examples that seem to confirm that Wells unfortunately completely mis-interpreted his teacher's personality. First, in the Christmas issue of *Nature* in 1879, writing under the pseudonym A. von Nudeln (*i.e.* Mr Macaroni), Guthrie humourously ridiculed a recent spate of German writings on geometry and mathematical physics in a letter entitled "On the Potential Dimensions of Differential Energy".

"In his great work, which appears to be but little known in England, *Über die stille Bewegegung hypotetscher Körper*, Professor Hans points out that the dimensions of 'ideal' matter may not only differ in degree, but also in kind. He deduced by means of implicit reasoning from his three primitive 'stations', that not only must there be space of 4, 5, 6, *etc.* dimensions, but also that there must be space of -1, -3, -5, *etc.* and that there may be space of -2, -4, -6 *etc.* dimensions. Pursuing Hans's train of thought further, Lobwirmski has quite recently interpreted space of 1.1, 1.2, 1.3 *etc.* dimensions ... the same philosopher has also conclusively shown that space of $n\sqrt{-1}$ exists [with] all the properties of angular magnitude; like all partly bounded infinities (*theilweise begränzte Unendlichkeiten*), it is unmagnifiable."

This piece of drollery ends with an obscure proof that the moon is, indeed, made of cheese. Guthrie's mixture of simplicity and wisdom, kindliness dashed with a pungent, but never caustic, humour, led one friend (perhaps Besant) to compare him with the Uncle Toby immortalized by Lawrence Sterne in *Tristram Shandy*.

This spoof probably supports Foster's contention in an obituary that Guthrie, though trained by De Morgan in mathematics, "was somewhat apt to underrate the scientific importance of the work of mathematical physicists in comparison with that of pure experi-mentalists". Guthrie's concentration on the physics of real, everyday things like water drops and bubbles, thermal conductivity of liquids, salt solutions, and the discovery of "cryohydrates" and the melting points of eutectic mixtures, and his use of a very informal homespun language to describe nature all seems to have been anathema to James Clerk Maxwell. Although Guthrie was elected to the Royal Society in 1873, as a referee Maxwell continually found fault with papers that Guthrie submitted to the Society, and this was undoubtedly the principal reason why Guthrie founded the Physical Society at the end of 1873.

It is again a remarkable example of Guthrie's ability to deflect criticism that in 1878 he was able to reply jokingly to Maxwell's intemperate review of his textbook, *Practical Physics, Molecular Physics and Sound*. Maxwell's particular gripe seems to have been Guthrie's language of exposition. For example, he picked on the way Guthrie formulated a variation of Hooke's law concerning the elongation of a wire, so that if its length m is extended to become n, the original diameter becomes $d\sqrt{m}/n$. The formula is correct, so presumably Maxwell thought it odd for Guthrie to say that the relationship is true only if one assumes "the volume of the metal remains approximately unchanged". Guthrie knew that Maxwell frequently wrote light verse, so his revenge was a splendid verse *à la Maxwell* written in Scots dialect, "Remonstrance to a Respected daddie anent his loss of temper".

WORRY, through duties Academic,
It might ha'e been
That made ye write your last polemic
Sae unco keen:
Or intellectual indigestion
O' mental meat,
Striving to solve some question
Fro' "Maxwell's Heat"....

This was signed $D\sqrt{m}/n$.

Following the 1867 Paris Exhibition and the success of the electric telegraph, the scientific community was able to promote the teaching of precision measurement in physics laboratories on the assumption that "hands on" experience was better than teacher demonstration if precision was to be achieved. In the first place, this would be good for encouraging rational and accurate reasoning within liberal education, and in the second, it would be a scientific alternative to "rule of thumb" methods of industrial apprenticeship, the need for which the electric telegraph had exposed. The problem was how to put this into effect. There were three common solutions put into practice in the 1870s.

In the first, the chemistry model, every student would have had a similar collection of apparatus. While this worked cheaply and well in chemistry, where practical work depended on little more than a set of test tubes, some filter and litmus paper and a bench stocked with standard chemicals, it would have been expensive to achieve the same goal in elementary physics teaching where relatively expensive optical and electrical equipment was necessary. Two alternative solutions to the problem of expense were therefore tried during the 1870s: one emanated from the Massachusetts Institute of Technology (MIT) and spread to Europe; the other was developed by Guthrie at South Kensington.

One of the more remarkable instructors at MIT in the 1870s was Edward Charles Pickering (1846–1919), a graduate from the Lawrence Science School at Harvard. William Barton Rogers, MIT's Principal, who thought that the Lawrence School was taking the wrong approach to science teaching, and believing that America desperately needed practical astronomers and physicists rather than engineers, appointed Pickering to MIT with the specific instruction to introduce practical physics teaching into the curriculum. Encouraged by Rogers, in the autumn of 1868 Pickering fitted up a student laboratory in Boston. With no European model to draw upon, and ignoring any chemical precedent, Pickering devised his own set of physics experiments, apparatus and instructions for students to follow.

In Pickering's system of instruction, which was ultimately cheap to operate and destined to become universally adopted in secondary schools and universities, students worked singularly or in pairs at different experiments, and proceeded in rotation from one experimental table to the next. Only one set of experimental apparatus was therefore needed per student pair. At each bench point students received written instructions on what to do, record and interpret. These experiments on techniques of measurement, the properties of gases, sound, the mechanics of solids and the nature of light (including photography), having been tested out at MIT for some three or four years, were published by Pickering in 1873 as *Elements of Physical Manipulation*. A second volume, largely devoted to electricity ("a subject better adapted than any other to the laboratory system") appeared in 1876 and included an appendix which offered advice on the planning of physical laboratories.

Not surprisingly in view of the fact that Pickering's text was reviewed in Norman Lockyer's *Nature*, Lockyer gave publicity to the MIT programme in the Sixth Report of the Devonshire Commission on Scientific Instruction, of which he was secretary. Knowledge of Pickering's method of multiple experiments was picked up quite quickly in Britain, probably from both the *Nature* review and from the Devonshire Commission. For example, in an essay on the teaching of physics published by Foster in the *Educational Times*, he referred approvingly to the "American" method of multiple classroom experiments.

It seems clear, then, that for anyone planning to develop a new physics teaching laboratory in the 1870s, the proven experience of the MIT laboratory would have been a good model to copy. The fact that William Ayrton, the electrical engineer, left London to teach at the Royal Engineering College in Tokyo in 1873, only a few months after the appearance in Boston of the first volume of Pickering's treatise, is suggestive. For it was in Japan that Ayrton and his mechanical engineering and mathematical colleague, John Perry, put the multiple experiment model of teaching into great effect. And on their return to London, they made this method of practical instruction famous at the Finsbury Technical College, from where it was adopted by the Cavendish Laboratory and other physics teaching institutions.

Although this method of instruction was destined to become universal, and is to this day the preferred method of teaching practical physics, there would have been initially high start-up costs. Indeed, until such times as instrument makers were geared up to mass-producing sturdy, cheap but reliable, Wheatstone bridges, galvanometers, lens systems, *etc.*, Guthrie's alternative approach must have seemed ingenious.

In Guthrie's system, derided and detested by H. G. Wells, but much admired by *Nature*, students were required to make their own apparatus, before manipulating it experimentally. The principal drawback, of course, was that the instruction method was time consuming. Guthrie gave a long account of the method and the educational philosophy behind it in his Cantor Lectures to the

Society of Arts in 1885. A decade earlier, in 1875, either he or the physicist William Barrett, informed *Nature*'s readers of the method's educational advantages.

> "Students unaccustomed to manipulation find to their astonishment, when they begin, that all their fingers have turned into thumbs, and they are amazed at their clumsiness and stupidity. Very soon, however, fingers begin to reappear, and the very first successful piece of apparatus that is made gives them a confidence in themselves which they had thought impossible to attain. The pleasure of having made an instrument is increased a hundredfold when it is found that by their own handiwork they may verify some of the more important laws of physics."

Since the final examination depended entirely upon the accuracy and reliability of the apparatus constructed by the student, there was an additional incentive to manipulate with care and attention.

It must not be thought that students were literally expected to make apparatus from scratch. Guthrie's purpose was certainly not to train and produce philosophical instrument makers. Apart from needing their own set of tools (a hammer, nails and other basic tools) in practice they received a kit which Guthrie's workman had already prepared and which they individually assembled by following printed instructions. For example:

> "[The student] received the wood and metal ... cut in pieces of the right size. He must have to acquire a little skill in bending and blowing glass, and in the use of the soldering iron. But what human being should be without this?"

The cost of materials for making the apparatus was some £2, compared with the cost of purchasing sets from a supplier for about £15—a saving to the Department of Science and Art of £13. Moreover, the student:

> "finds himself at the end of the course with a set of apparatus made by himself, and fully tested by himself; and he finds himself in possession of verifications of many of the great

generalizations of physics, which generalizations, having been thus acquired by the simultaneous working of brain and hand, he is slow to forget."

Although Guthrie did not mention it, the method also had an advantage over Pickering's system insofar as a newly certificated science teacher from South Kensington could bring his own set of apparatus with him to his new school teaching post. This would have been important where finances were tight and a new teacher had first to consolidate his position in a school before he could argue for the purchase of apparatus from instrument suppliers.

Finally, we should note the wider influence of Guthrie on physics teaching in heat, light, sound, electricity and magnetism. Just as Edward Frankland exploited his examination position to lay down a chemistry syllabus and to demand that all students either witnessed or conducted a definite number of experiments, so did Guthrie. In 1881 Guthrie's *Outline of Experiments and Description of Apparatus and Material Suitable for Illustrating Elementary Instruction in Sound, Light, Heat, Magnetism and Electricity* was published by the Department of Science and Art for science teachers. These recommendations not only influenced physics teaching in Britain, but also in Japan and America. Kannosuke Yoshioka, the Japanese translator of Guthrie's *Practical Physics* (1877), incorporated Guthrie's recommendations and also arranged for a set of apparatus to be displayed at the Tokyo Education Museum from 1895 onwards. It appears, then, that Guthrie played a significant posthumous role in rendering Japanese science teaching less didactic and more practical.

In America, there was little "hands on" experimental work by High School students until the 1890s. Dr Alfred Gage, who opened an English (*sic*) High School in Boston in 1880, and who published a practical textbook, *Elements of Physics* in 1882, significantly abandoned teaching in order to make and supply scientific instruments for school use. In the same decade, Frank Wigglesworth Clarke's report on the state of physics and chemistry teaching in American city schools showed that less than ten per cent possessed laboratories. This situation began to change during the 1880s when Charles William Eliot, President of Harvard University, and himself a trained chemist, deliberately placed physics on the list of matriculation requirements.

Initially, in 1886, Harvard allowed two alternatives—either a written examination based upon the use of designated textbooks of astronomy and physics, or a practical demonstration based upon the matriculant's school experience of a "course of experiments in the subject of mechanics, sound, light, heat and electricity, not less than 40 in number". Although these 40 experiments were chosen from American texts like Pickering's *Physical Manipulation*, the choice was also influenced by British texts such as Guthrie's *Practical Physics* and A. M. Worthington's *Physical Laboratory Practice* (1886). By the 1890s, Harvard insisted upon evidence of practical experience of its physics for its entrants into science courses. We may, therefore, conclude that, directly or indirectly, Guthrie also influenced the rise of the elementary physics laboratory in America.

By way of conclusion, I draw attention to the organic chemist Henry Edward Armstrong who, writing in 1933, a year before Wells's cruel portrait of Guthrie appeared in his autobiography, gave a Huxley Memorial Lecture at South Kensington. In this Armstrong makes a sudden aside:

"Good as was the top floor [of the Normal School where Huxley taught], it was being beaten, down below in the basement, by the chemist-turned-physicist, Guthrie, who developed a logical, practical course, on self-help lines, of extraordinary value—long since on the scrapheap, I fear, given its final quietus by the all-pervasive electron. It is little short of shameful that South Kensington has let this go by the board unrecorded. Real earthly physics is a fast-disappearing art. Cannot someone be found to recover this course, if only to put it away in a case in the British Museum, as a monument of a former greatness?"

When Foster and Guthrie are rescued from the scrapbook of history, we can see the significance of their roles and that of other chemists in the establishment of practical physics teaching.

CHAPTER 13

Chemical Algebra

In 1834 Thomas Graham, the future first President of the London (now Royal) Chemical Society, applied for the chair of chemistry at University College London. John Dalton, the founder of the chemical atomic theory, was one of Graham's referees. Dalton's curious testimonial tempered the expected praise with criticism of the fact that Graham had recently used chemical formulae and equations in a paper published in the Royal Society's *Philosophical Transactions*. "Berzelius's symbols are horrifying", Dalton complained. "A young student in chemistry might as soon learn Hebrew as to make himself acquainted with them. They appear like a chaos of atoms. They equally perplex the adepts of the science, discourage the learner, as well as cloud the beauty and simplicity of the Atomic Theory".

Symbols and a symbolic language mark the chemist from all of the other scientific disciplines apart from mathematics. While most outsiders know that H_2O means water, they would be defeated by CH_3COOH, and totally mystified by the more complex arrays of symbols chemists regularly use to portray molecules in three dimensions. Even the real meaning of H_2O, namely one molecule of water containing two atoms of hydrogen combined with one atom of oxygen, is likely to be beyond the powers of interpretation of the non-chemist. When and why did chemists adopt this symbolic language?

The Case of the Poisonous Socks: Tales from Chemistry
By William H. Brock
© William H. Brock, 2011
Published by the Royal Society of Chemistry, www.rsc.org

Jöns Berzelius, the Swedish pope of chemistry during the first four decades of the nineteenth century, first introduced the formulae we use today in 1813. They found little use beyond shorthand for the labelling of bottles or for listing atomic and equivalent weights. And despite Lavoisier's introduction of a stoichiometrically balanced verbal equation in one of his papers at the end of the eighteenth century, equations too also remained unexploited. As the German scholar Ursula Klein demonstrated in her monograph, *Experiments, Models, Paper Tools* (2003), all this changed in the 1830s when organic chemistry burst upon a chemical community that was already conscious of the quantitative usefulness of Dalton's atomic theory. The catalysts were the discovery of isomerism (in which two or more substances containing exactly the same amounts of different elements have quite different chemical properties) and Justus Liebig's wonderful development of a simple, rapid, elegant and reliable method of organic analysis by the direct weighing of combustion products. Both these events occurred around the year 1830. The consequence was an explosion in the number of artificial carbon compounds (as opposed to the natural products that had formerly been the sole concern of agricultural and physiological chemists in what was referred to as vegetable and animal chemistry). How could one make sense of so many new derivatives and isomers that did not exist in the natural world? How should they be classified?

The answer was to exploit Dalton's and Berzelius's atomic weights in combination with chemical formulae to develop a "paper chemistry" in which plausible relationships between stoichiometrically valid reactants and products were sketched out on paper in the form of equations. Such paper tools and paper trails were first exploited independently by Jean Baptiste Dumas in Paris and Justus Liebig in Giessen. Their students and contemporaries followed suit, so that formulae became another laboratory tool. The exploitation of the "graphic suggestiveness" of Berzelian formulae for representing organic reactions, and for constructing models of their composition, was to lead chemists along the pathway to structural theory in the 1860s and to three-dimensional models in the 1870s. It also enabled them to classify substances into acids, ethers, paraffins, *etc.*, and so bring the overwhelming number of organic substances into some kind of order. For example, in 1832, Liebig and Wöhler separated and identified nine reaction products of the oil of bitter almonds (none of which had been present in the original nuts). They then interpreted the products on paper as binary compounds of a benzoyl radical ($C^{14}H^{10}O^2$) combined with a single element such as H, O or Cl, or a compound like ammonia. Oil of bitter almonds was therefore "benzoyl hydride". Subscript formulae were adopted in the 1840s.

Although unitary models that involved substitution and addition soon superseded this radical dualism model, the method of paper chemistry caught fire. Berzelian signs, together with balanced stoichiometric equations were found to aid the design of experiments, modelling, theorizing and classification—all necessary tools for navigating through what Wöhler had called the "dark forest" of organic chemistry. Formulae and equations filled in the gaps left open ended by experiment, as was shown erroneously (but brilliantly) by Liebig in his *Animal Chemistry* in 1840. There he "invented" the metabolic pathways between the empirically determined constituents of air and food intakes, excretion products and natural products found in flesh and blood to create plausible paper biochemistry (Figure 13.1).

It is easy to see why a committee of the British Association for the Advancement of Science rejected Dalton's atomic hieroglyphics for Berzelian alphabetical signs in 1834. Agreement among chemists on the exact form Berzelius's system of signs should take was, however, a matter of negotiation. The Cambridge mineralogist and

Figure 13.1   A chart dating from 1808 illustrating the chemical symbols of the elements used by John Dalton. To use more generally special type would have had to be made by printers. In comparison, ordinary typefaces could be used for Berzelius's symbols. (Science & Technology Picture Library)

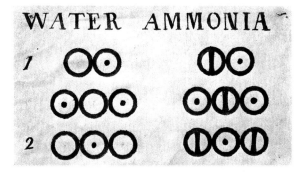

Figure 13.2 Dalton's chemical symbols applied to compounds. (Wellcome Images)

polymath William Whewell, for example, wanted Berzelius's symbols to be written in a way that was truly algebraic. The compound A^nB^n should be formulated as $nA + nB$, with lots of brackets for more complex molecules. Other, more pragmatic chemists soon objected that this was quite unnecessary and made matters too complicated. The index figures in Berzelian formulae did not represent algebraic powers, just as the juxtaposition of symbols did not denote multiplication. In arithmetic, everyone knew (by convention) that CLV in Roman numerals, or 155 in Hindu numerals, was a decimal system and not to be interpreted as $C \times L \times V$ or $1 \times 5 \times 5$. Similarly A^n in chemistry did not mean $A \times A \times A \dots n$ times, but simply n atoms of the element A. Whewell sensibly withdrew his suggestion (Figure 13.2).

Compared to Dalton's typographical phantasmagoria of geometrical circles, algebra-like signs using the standard alphabet were easier to print and write, they had constitutional or structural implications just like algebraic factors, and relationships between reactants and products could be easily seen when formulae were expressed in equation form. Graham got his job at University College London. The twenty-first century chemist, faced by the fleeting quasi-molecular species revealed by spectroscopy, would find it impossible to work without a symbolic system, albeit the laboratory tools of paper and blackboard have now been extended to, and largely replaced by, computer modelling.

CHAPTER 14

The B-Club

The B-Club (or Hive Club) flourished between 1860 and 1875. It was a convivial dining and drinking club of London chemists and their friends that took its title from Section B, the chemistry section, of the British Association for the Advancement of Science (BAAS). Many of its members were former students of A. W. Hofmann at the Royal College of Chemistry (RCC), while others were connected through William Francis, a trained chemist who co-edited and printed the monthly *Philosophical Magazine* and had interests that lay outside chemistry.

The club was modelled on the Red Lion Club founded by the naturalist Edward Forbes at the BAAS meeting at Birmingham in 1839. This convivial drinking and dining club met annually at provincial meeting of the BAAS. From 1841 onwards opportunities for the discussion and promotion of both chemistry and pharmacy had been helped by the foundations of the Chemical and Pharmaceutical Societies. Friendships among London chemists and pharmacists, especially among the several chemists who worked in London's hospitals, probably suggested the creation of a less formal and more convivial meeting place than the societies' premises.

The B-Club was founded in about 1860 by the hospital chemist William James Russell. It met monthly at the Rainbow Tavern in Fleet Street for informal dinners. Formerly one of London's first seventeenth-century coffee houses, the Rainbow was famous for the quality and quantity of its stout, as well as for its beef and

The Case of the Poisonous Socks: Tales from Chemistry
By William H. Brock
© William H. Brock, 2011
Published by the Royal Society of Chemistry, www.rsc.org

mutton meals. Francis printed the invitations using notepaper headed by drawings of three bees and replies were sent to Russell. There was also an annual summer excursion into the countryside. Although limited to twenty members in any one year, guests regularly swelled the club's numbers. Discussion of chemical matters was light-hearted and the members' drinking songs and buffoonery was greatly aided by the literary and artistic talents of the dyestuffs chemist Frederick Field, the pharmaceutical journalist John Cargill Brough, the organic chemist and physicist Frederick Guthrie, and the lawyer William Phipson Beale, who had worked as a chemist in his uncle's steel works in Sheffield before being called to the bar. Although he practised as a lawyer for the remainder of his life, Beale continued to take an interest in chemistry and mineralogy. He published a book on crystallography in 1915 at the age of 76 (Figure 14.1).

The witty drawings that embellished menu cards and the satirical songs and ballads were the great attractions of the club's dinners.

Figure 14.1   John Cargill Brough (1834–1872), a pharmaceutical chemist and leading member of the Chemical Society's B-Club. Brough edited the monthly *Chemist and Druggist* from 1859. (Brock)

During the 1860s these songs lampooned debates over structural formulae, the safe disposal of sewage, and court cases that involved disputes over patents for the new synthetic dyestuffs. In about 1875 the club merged into the Chemical Club organized by Henry Edward Armstrong at that time.

Here are two verses from Forbes and Price's *On the Utilization of Sewage by Phosphate of Alumina. A Paper which ought to be sung by the B-Club.*

> There is not in the wide world a water so sweet
> As the effluent liquid when sewage we treat:
> Oh! The last trace of organic foulness departs
> With the charm our alumina imparts.
>
> Sweet alumina phosphate, how calm we could rest
> By the side of a drain with the drink we love best,
> If Bazelgette, Hawksley and Denton would cease,
> And our stuff with the sewage were mingled in peace.

The last stanza was an obvious reference to contemporary discussions and controversies over the plans of civil engineers to rid Londoners of the menace of urban sewage through costly engineering works. Chemists like Forbes and Price believed that town sewage could be rendered harmless much more cheaply by using chemicals (like aluminium phosphate) to precipitate and de-odorize the sewage.

Three stanzas from Brough's *Modern Chemistry*, which was dedicated to the explosives chemist Frederick Abel in March 1868, brilliantly recreated the bewilderment of many chemists faced by the doctrine of valence and its disconcerting symbolism developed by William Odling, Edward Frankland and Alexander Crum Brown, and forced upon students through the government's examination systems administered by Kay Shuttleworth.

> I'm all in a flutter; I scarcely dare utter
> The words I have set to a jingle;
> For I see at this table philosophers able,
> Whose ears at my verses will tingle.

Still, I don't mind confessing, I'm fond of expressing
My notions and thoughts, in defiance
Of every great gun who can't see the fun
Of winnowing chaff out of science.

I've read till I'm weary books weighty and dreary
In which certain chemists seem aiming
To prove to outsiders they're excellent riders
Of hobbies in theory and naming.
With "monads" and "diads", and "pentads" and "triads",
My brain has been addled completely;
And what's really meant by "something-valent",
Is a question I give up completely.

Though Frankland's notation commands admiration,
As something exceedingly clever,
And Kay Shuttleworth praises its subtle worth,
I give it up sadly for ever:
Its brackets and braces, and dashes and spaces,
And letters decreased and augmented
Are grimly suggestive of lunes to make restive
A chemical printer demented.

I've tried hard, but vainly, to realize plainly
Those bonds of atomic connexion,
Which Crum Brown's clear vision discerns with precision
Projecting in every direction.
In fine, I'm confounded with doctrine expounded
By writers on chemical statics,
Whom jokers unruly may designate truly
As modern atomic fanatics.

Further verses lashed out at Benjamin Brodie's Greek alphabetical symbolism which attempted to replace the familiar Berzelian atomic symbols, expressed bewilderment at the latest ideas on acidity, and at Odling's strange terminology of "aplones" and "diamerones".

Historians of science now recognize that jocularity and playfulness have social roles to play in the organization and practice of science. Such jesteral approaches seem to have played a particularly important role in the history of chemistry—witness the ritual of laughing gas extravaganzas that were often permitted annually in student laboratories and the tradition of annual "joke" chemistry lectures to students and staff that lampoon staff, rival departments or the current chemical literature.

Both the B-Club and the later Chemistry Club deliberately limited their membership—no doubt for the practical reason of sitting comfortably around one dining table. However, to some outsiders the members seemed like a clique and were often condemned as "the pot-house brigades". The solution to the problem of sustaining and improving social relations among chemists was to hold regular soirées for members and their wives. These were started perfunctorily in 1878 by John Hall Gladstone and regularized in 1883 by Willliam Henry Perkin. The soirées took the form of joint gatherings of the Chemical Society, the Society for Chemical Industry and the Institute of Chemistry at the Princes Hall in Piccadilly. Annual dinners were also instituted in 1890. The communal events continue today in the form of an annual reception each summer at the Royal Academy of Arts when chemists mix conversation with drinks and looking at works of art.

Known Members of the B-Club

Charles D. Abel

Frederick A. Abel (1827–1902), RCC, explosives chemist at Woolwich; a skilled amateur pianist and singer

Edmund Atkinson (1831–1900), pupil of Robert Bunsen, he taught natural philosophy (chemistry and physics) at Sandhurst. A friend of Francis and Hirst

Sir William Phipson Beale (1839–1922), lawyer and later an MP, author of *Amateur's Introduction to Crystallography* (1915)

Henry Bessemer (1813–1898), bronze powder manufacturer and metallurgist

E. F. Best

John Cargill Brough (1834–1872), editor of *Chemist & Druggist* and *Laboratory*

George Bowdler Buckton (1818–1905), RCC, entomologist, Odling's brother-in-law and friend of Francis and Van Voorst

Baldwin Francis Duppa (1828–1873), RCC; assisted W. H. Perkin and Edward Frankland

Frederick Field (1826–1885), RCC, dyestuffs chemist at Simpson, Maule & Nicholson and Abel's brother-in-law

David Forbes (1828–1876), geological and mining chemist and younger brother of Edward Forbes; patented a sewage deodorizer with Price

George Carey Foster (1835–1919), organic chemist and Professor of Physics at University College from 1865

Michael Foster (1836–1907), doctor and physiologist, later Professor of Physiology at Cambridge

William Francis (1817–1904) chemist, entomologist, printer and publisher of the monthly *Philosophical Magazine*; former editor of the *Chemical Gazette*

Frederick Guthrie (1833–1886), organic chemist, later Professor of Physics at the Royal School of Science

Thomas Archer Hirst (1830–1892), geometer, friend of Atkinson and Francis

M. Holzmann, a former student of Kekulé at Ghent

John Lubbock (1834–1913), banker and naturalist

Augustus Matthiessen (1831–1870), pupil of Robert Bunsen, hospital chemist and electrician

William Allen Miller 1817–70), professor of chemistry at King's College, London

Hugo Müller (1833–1915), pupil of Friedrich Wöhler, dyestuffs consultant and De La Rue's chemical assistant

Edward C. Nicholson (1827–1890), RCC, dyestuffs chemist and partner in firm of Simpson, Maule & Nicholson

Henry M. Noad (1815–1877), RCC, chemist and electrician at St George's Hospital

William Odling (1829–1921), RCC, organic chemist

Astley Paston Price, patentee with Forbes of aluminium phosphate deodorizer

Trenham Reeks, registrar and curator of the Royal School of Mines

Warren de la Rue (1815–1889), RCC, chemist and astronomer, as well as printer and paper manufacturer

William James Russell (1830–1909), physical chemist at Bedford College for Women

Alfred Smee (1818–1877), surgeon, electrometallurgist, and Odling's father-in-law

John Van Voorst (1804–1898), natural history publisher

CHAPTER 15

Chemistry by Discovery in a Phrase

The slogan "I hear, I forget; I read, I remember; I do, I understand", or variations of it, is frequently heard at education conferences or even seen emblazoned on T-shirts. The message is clear: teachers want science to be a "hands on" experience for their pupils. But who first coined the aphorism?

The most obvious source is the English organic chemist Henry Edward Armstrong (1848–1937) who campaigned tirelessly for practical teaching of science from the 1880s onwards until his death. He described this method of teaching science as "heuristic", echoing Archimedes's "Eureka" cry when the idea of specific gravity came to him in the bath. However, although the slogan epitomized the heuristic method that Armstrong campaigned for, it will not be found in his extensive educational writings (though there are phrases about "doing to understand").

On the other hand, Armstrong was very successful in getting heurism adopted in the famous charity school Christ's Hospital when it transferred from the City of London to Horsham in Sussex at the beginning of the twentieth century. It must have been one of the brilliant science teachers at the school who first coined the phrase. The clue comes in a little book, *The Teaching of Science at Christ's Hospital since 1900 A.D.*, which was privately printed by Gordon Van Praagh in 1992. Apart from a break doing research for the Admiralty at Imperial College, London, during the Second World War, Van Praagh (1909–2003) taught chemistry at Christ's

The Case of the Poisonous Socks: Tales from Chemistry
By William H. Brock
© William H. Brock, 2011
Published by the Royal Society of Chemistry, www.rsc.org

Hospital from 1934 until 1964. In 1949 he published a revolutionary laboratory textbook, *Chemistry by Discovery*, in which a pupil was led, as far as is practicable, to "discover chemistry for himself". For example, the pupil would heat a sample of copper. Why did it go black? Had it burnt like toffee? Had soot come from the flame? Had an impurity in the metal been forced out by heat? Had the air caused it? By devising tests for these hypotheses, the pupil was led to the conclusion that air was responsible and so to new investigations concerning what part of air was responsible for combustion. Such a procedure involved active thinking by pupils rather than their being at the receiving end of information and lecture demonstrations about the role of oxygen in chemical change. In true Armstrong spirit, VP (as he was known in the school) did not reveal the outcome of any experiment explicitly, though he assumed the result in subsequent sections of the book. It proved a milestone in the spread of teaching by modified heuristic method that he called "guided discovery" both in the UK and overseas.

The publication of *Chemistry by Discovery* led to Van Praagh's recruitment to the syllabus revision committee of the Science Masters' Association in 1960. He then joined the chemistry team of the Nuffield School Science Project, continuing its work abroad through the British Council and the Ministry of Overseas Development. Writing about his predecessor as a science teacher at Horsham, Praagh says:

"What a pupil finds out for himself, either by direct experience or through the pages of a book, he not only remembers better but understands more clearly. 'I hear and forget,/ I see and remember,/ I do and understand.' This saying, derived from similar sentiments expressed in the writings of Chinese philosophers, was adopted by the Nuffield Project as a sort of theme song."

The proverbial triad is, in fact, attributed to Confucius. Presumably it (or its variant) also caught on in North America among curriculum reformers in the same decades.

Van Praagh adds an amusing example of the need for discovery methods from Samuel Pepys's diary entry: "This day I did pay my admission money to the Royal Society. Here was very fine discourses and experiments, but I do lack the philosophy enough to understand them, and so cannot remember them".

Part 3: A Cluster of Chemists

From time to time members of the public are asked to name a famous scientist. Predictably the answers are restricted to a handful of names such as Newton, Darwin and Einstein. Perhaps Faraday's name will be recalled, but although Faraday trained as a chemist, the public will undoubtedly be imagining him as an electrician or physicist. Where are the unsung heroes of chemistry? The answer would appear to be that because chemistry is the basic science that underwrites all of the other sciences, chemists' contributions have been and tend to be more associated with them rather than with chemistry *per se*. To help restore the balance here are a random sample of nine individual chemists who have made significant contributions to the science through their teaching, research, or criticism.

We begin with the Italian, Avogadro, who underlines again the important role that standardization plays in science. Although the clarification his work offered was ignored for several decades, it finally turned out to provide a way out of the anarchy generated by the rival systems of molecular classification that had been developed by different schools of chemistry. An illustration of this anarchy is given in the story of the collaboration of Wöhler and Liebig (collaboration being itself a new phenomenon). Although their particular interpretation of their preparative work in organic chemistry (the radical theory) was soon overturned, it was an important step in providing enlightenment during an otherwise bewildering period in the history of chemistry. The final clarification of organic chemistry owed much to the authority of Kekulé whose shrewd theoretical insights provided chemists with the fundamental building blocks of structural chemistry, namely the carbon chain and the benzene hexagon.

The fourth and fifth stories are concerned with iconoclasm. The case of Brodie reminds us that the validity of an atomic theory of matter was by no means accepted by the nineteenth-century scientific community. Brodie, agnostic in both religion and science, was prepared to abandon atomism completely and to develop an alternative system of frightening complexity. As a thought experiment, it would be an amusing exercise to imagine contemporary chemists using his calculus to interpret Buckminster fullerenes. Another organic chemist, Henry Armstrong, was also worried about chemists' increasing reliance upon theory. He urged strongly that chemists should never stray from an empirical base. Despite his varied contributions to organic chemistry and to the organization of the Chemical Society, he has become better remembered as a critic. But iconoclasts and critics like Brodie and Armstrong, who may end up in dead ends, undoubtedly play an important role in stimulating logical thought. Tilting at windmills can advance chemistry.

The tragic story of Fritz Haber may be well known, but it is worth retelling because it highlights one of the biggest problems chemists face in the eyes of society—the fact that "chemicals" have the power to both improve and harm the human condition. One of the chemists drafted to examine how German chemists had developed the Haber process for the fixation of nitrogen during the First World War was J. R. Partington. His career exemplifies the ordinariness of most chemists' lives. Partington was the typical academic chemist who conscientiously teaches undergraduates, trains postgraduates, publishes large numbers of papers and textbooks, but who makes no outstanding discoveries. Such lives should not be denigrated or ignored. In fact, although Partington typifies the life of a twentieth-century chemist, he was atypical in the prolific and encyclopedic nature of his publications and in the fact that he became better known as an historian of chemistry than as a practising chemist.

Although there have been several chemical dynasties like the Gmelins, the Franklands and the Perkins, it has always been difficult for the sons and daughters of famous chemists to make names for themselves or remain on good terms with their parents. In more recent times one thinks of Peter Pauling, Roger Partington and Keith Ingold. This phenomenon is illustrated in the penultimate story about Henry Crookes. His tragedy was that he died at the very moment that his enterprise to make colloid chemistry commercial became successful. It was left to others to develop the pharmaceutical company that bore his name, or was it, even more ironically, the name of his father?

The final story briefly concerns one of the great debates of twentieth-century chemistry. Could one demonstrate the existence of unstable intermediates in chemical reactions? Or were they just like Brodie's atoms, figments of the imagination? The case of George Olah also reminds us of the expensive nature of modern chemical research. Gone are the days when research could be financed by students' fees topped up by government grants. The twenty-first century chemist is now dependent upon goal-driven industrial, military and pharmaceutical funding and he or she must be as astute in grantsmanship as in laboratory manipulation.

CHAPTER 16

Amedeo Avogadro

When the American Chemical Society celebrated its centenary in 1974 it issued car stickers that proclaimed the message "I know Avogadro's number is 6.022×10^{23}". The implication was that, like C. P. Snow's claim for knowledge of the Second Law of Thermo-dynamics as a mark of good education, familiarity with Avoga-dro's number was a test of the divide between science and the humanities. On the other hand, the Society may have meant that knowledge of the number marked the chemist from the physicist. The irony is that the number was never calculated by the Italian chemical physicist and the eponym was only made fifty years after his death. The eponym is justified on the grounds that Avogadro was one of the founders of modern molecular theory and of the standardization of atomic weights (Figure 16.1).

The splendidly named Lorenzo Romano Amedeo Carlo Avogadro, the Count of Quaregna and Cerreto, was born in the Piedmontese city of Turin in the kingdom of Sardinia on 9 August 1776, the son of a prominent civil servant who was charged under the Napoleonic rule of 1799 to reorganize the Piedmont govern-ment. The family had a long tradition of legal service to church and state—indeed, the surname Avogadro is a corruption of the Latin *advocato*, a barrister. Avogadro graduated in law from the University of Turin in 1795, and was awarded a doctorate in Canon Law the following year. He then followed in his father's footsteps

The Case of the Poisonous Socks: Tales from Chemistry
By William H. Brock
© William H. Brock, 2011
Published by the Royal Society of Chemistry, www.rsc.org

Figure 16.1 Amedeo Avogadro, Conte de Quaregna (1776–1850), Turinese
physical chemist who suggested in 1811 that equal volumes of
gases at the same temperature and pressure would contain the
same number of molecules. (Brock)

by working for the government, changing administrative roles after
the French invaded Piedmont in 1799. Like Antoine Lavoisier, who
had also studied law, Avogadro began to develop scientific interests
in his spare time. He was particularly stimulated by the excitement
generated by Alessandro Volta's development of the electric battery in 1800. Interest in electricity prompted him to audit a course
of lectures and demonstrations on physics at the University of
Turin and to join the local Academy of Sciences, to whom he
submitted his first two papers in 1804. The subjects of these papers,
the nature of insulators (dielectrics) and the electromotive series,
remained dominant interests of Avogadro throughout his life. The
application of mathematics to physical observations in an attempt
to understand and classify chemical relationships between elements
and compounds (acidities, affinities, atomic volumes and specific
heats) was to mark Avogadro out from contemporary chemists,
who largely ignored his publications. Unlike his fellow countryman
Volta, Avogadro was indifferent towards a reputation outside
his own country, which he never left to meet other natural

philosophers. A theoretician in an age of experimentalists, and hindered by the political situation in Piedmont, Avogadro's work, which was often out of kilter with what was going on in the rest of European science, was destined for neglect in his lifetime.

In 1806, determined to help reverse his country's perceived backwardness in the sciences, Avogadro abandoned his legal career and began to teach mathematics and physics at a succession of Piedmontese schools and colleges until, in 1809, he was appointed Professor of Natural Philosophy at the former Royal College in Vercelli, 60 km east of Turin. Living under a French political regime, he was inevitably attracted to French scientific culture and so began sending his scientific papers to Jean-Claude de Lamétherie's monthly *Journal de physique*.

Traditionally historians of chemistry have seen Avogadro as proposing his molecular hypothesis to reconcile John Dalton's suggestion that atoms combine gravimetrically in the simplest possible way, as when single atoms of hydrogen and oxygen combine to form one compound atom of water:

$$H + O = HO$$

and Joseph Gay-Lussac's 1809 discovery of the integral law of combining volumes of gases:

$$2 \text{ volumes } H + 1 \text{ volume } O = 2 \text{ volumes water}$$

(Berzelian symbols, which Avogadro did not adopt, are used for clarity.)

Gay-Lussac's experiments suggested that equal volumes of different gases ought to contain the same number of ponderable particles, provided that the physical conditions were kept constant. Yet, in practice, there was a glaring discrepancy between the expected ratio of volumes (in the case of water, 2 : 1 : 3) and the observed ratios (2 : 1 : 2). Moreover, the densities of the gaseous products were frequently at odds with those of the reactants. For example, in the case of the oxidation of carbon monoxide to form carbon dioxide:

$$100 \text{ volumes } CO + 50 \text{ volumes } O = 100 \text{ volumes } CO_2$$

the density of carbon dioxide (1.5196) was *less* than that of the half unit of oxygen (0.5518) that had gone into its formation. How could a gas that contained both carbon and oxygen be lighter than oxygen itself?

The gravimetric and volumetric information would make sense, Avogadro suggested in a French paper in 1811, if equal volumes of gases under the same physical conditions of temperature and pressure contained the same number of "integral molecules", the latter being separable aggregates of at least two, but possibly more, particles. In other words, he was suggesting that both elementary and compound gaseous molecules divide at the moment of their reaction with other elementary and compound molecules to form the observed volumes of the products. For example, water was formed from a half-molecule of oxygen and two half-molecules of hydrogen:

$$2H_2 + O_2 = 2H_2O$$

Because the numbers of such aggregated (or integral) particles were the same in equal volumes, it followed that the ratios of their densities (relative to a chosen standard density) stood in the same ratio as their relative masses.

In the four papers Avogadro published on the subject between 1811 and 1821 he suggested how, on the basis of this hypothesis, vapour density measurements might be used to determine molecular weights, assuming that oxygen of molecular weight 16 was at least a dimer. In later parlance, this gave the familiar formula: molecular weight = $2 \times$ vapour density. There was never any implication that the particles that made up integral molecules were physical atoms and Avogadro probably did not believe in the existence of such indivisible entities. He did not, therefore, as was assumed by later chemists, distinguish between atoms and molecules.

As the historian Nicholas Fisher showed in 1982, it is hardly surprising that Avogadro's suggestion was ignored or dismissed as fanciful speculation (though the Frenchman André-Marie Ampère did play with the notion in 1816). At the time the concept of two identical atoms of an element combining to form a stable dimer was inconceivable (and indeed remained mysterious until the emergence of the notion of electron sharing in the early 1900s). According to the chief chemical authorities of the day, John Dalton and Jöns Berzelius, like atoms were self-repulsive, because they were surrounded either by mutually repulsive atmospheres of heat (caloric) or of electricity. As an added twist, Avogadro's model had gas molecules surrounded by atmospheres of caloric, leading him to suppose they must be the same size in all gases. A further difficulty was that Avogadro could only determine the vapour densities of a handful of gases and his theoretical calculations of the vapour densities of solid elements worked to four or five decimal places, based upon the weight of the element that combined with 16 parts (one volume) of oxygen, seemed entirely hypothetical. Finally, readers of his papers were probably confused by his cumbersome usage of the terms *molecule* (either an atom or molecule), *integral molecule* (usually the molecule of a compound), *constituent molecule* (one molecule of the element) and *elementary molecule* (one atom). In fact this molecular vocabulary was unique to French writers at the end of the eighteenth century and provides historians with the clue to understanding Avogadro's research programme.

The geographically isolated Avogadro belonged intellectually to the French research school of Lavoisier's colleague Simon Laplace, the mathematician and astronomer, and Lavoisier's chemistry disciple, Claude Berthollet. In his famous dispute with Louis-Joseph Proust at the end of the eighteenth century Berthollet had insisted that constant composition was limited in application. Chemical composition was determined by physical conditions and concentration, since these affected the natural affinities between elements. Although Dalton had built the laws of constant and multiple composition into his atomic theory at the beginning of the nineteenth century, Berthollet's version of the law of mass action (as it became known in the 1870s) intrigued Avogadro who spent much of his life trying to determine absolute affinity figures from physical data in support of the French programme. Gases were of a primary interest to him (as they were to Berthollet's pupil Gay-Lussac) because, unlike solids and liquids, their affinities were completely saturated. Unfortunately, after a promising start, Berthollet's research programme, which involved the caloric theory as well as Laplace's idea of short range forces of attraction and repulsion, led nowhere and was abandoned by French scientists after about 1815. Avogadro, however, kept the faith; consequently his later work failed to enhance his stature among European chemists. His multi-volume physics textbook, *Fisica dei corpi ponderabili*, published when he was in his sixties between 1837 and 1841, showed he had been left behind completely by the advances in physics in other countries.

Avogadro remained at Vercelli until 1820 when he returned to the University of Turin as its first Professor of Mathematical Physics only to lose it a year later when political upheaval caused the chair to be suppressed. During the next ten years he returned to the practice of law, but was reappointed to the re-established Turin chair in 1834. In 1850 he and his wife retired to their family estates at Quaregna. He died on a visit to Turin on 9 July 1856. None of his obituaries mentioned the molecular hypothesis, though they described him as "religious without intolerance, learned without pedantry, wise without ostentation, a despiser of pomp, without care for riches, not ambitious for honours; ignorant of his own worth and fame, modest, temperate, and lovable". Although warmly regarded by the Piedemontese as a man of great learning and modesty, he had been little known outside the Italian peninsular during his lifetime, despite his French offerings. Avogadro's career supports the views of sociologists of science who stress that historical, social, political and cultural factors play important roles in the emergence of scientific knowledge. Avogadro sat at the edge of the French scientific empire but lacked any personal contact with French (let alone English and Scandinavian) chemists. The result was intellectual isolation and invisibility.

This isolation only altered with the *Risorgimento* that culminated in the unification of Italy in 1870. Italians wanted to underwrite their new nationhood by looking for cultural heroes. Posthumously Avogadro became seen as one of the founders of molecular theory after Stanislao Cannizzaro showed in 1858, and more forcibly at the Karlsruhe Congress of chemists two years later, how a rational and standardized system of atomic weights could be achieved by distinguishing between atoms and molecules, and placing Avogadro's hypothesis at the forefront of chemical theory. The different quantities of the same element contained in one molecule of the free substance and in those of its compounds were then always integral multiples of a quantity that represented the atomic weight. This standardization removed and rationalized a great deal of confusion and complemented the views of contemporary organic chemists who had been deeply ashamed of the muddle and confusion within their subject.

By the 1870s the kinetic theory of gases had given credence to the atomic–molecular theory and it was the Bohemian chemist Josef Loschmidt (1821–1895) who used the gas laws to calculate the number of molecules in a standard volume, the cubic centimetre. His value of 2.6×10^{19} has since been recalculated as $2.68 \times 10^{25}\,\mathrm{m}^{-3}$. This is still known as Loschmidt's number. However, when Jean Perrin used Brownian motion to provide convincing proof of the existence of molecules in 1909 he used the gram molecule (mole) as his standard and calculated the value as 6×10^{23}. Today this is defined as the number of molecules in 12 grams of the carbon 12 isotope. It was Perrin who suggested that this important chemical constant should be designated as Avogadro's number in commemoration of Avogadro's original insight in 1811.

As Bill Bryson notes in *A Short History of Everything*, American college students in the twentieth century estimated that the stupendous figure of 6×10^{23} was equivalent to the number of popcorn kernels needed to cover the United States to a depth of nine miles. Do Italian students have a pizza version?

Liebig and Wöhler: Creating a Path through the Dark Province of Organic Nature

With the sole exception of Liebig's pupil, the Anglo-German bio-chemist, John Thudichum, who ignored Wöhler's relationship with Liebig in his Cantor Lectures on Liebig's life given to the Society of Arts in London in 1875, all biographers of Liebig and Wöhler have commented upon their remarkable friendship and their fruitful and creative collaboration. Why Thudichum chose to ignore the friendship is not known. Wöhler was the great "all-rounder" of chemistry. He was above all a laboratory-bench, hands-on chemist who delighted in attacking any practical problem in mineralogy, inorganic or organic chemistry. He told Liebig in 1863:

> "My imagination is pretty active, but in thinking I am very slow. No one is less made a critic than I. The organ for philosophical thought I lack completely, as you well know, as completely as that for mathematics. Only for observing do I possess, or at least I believe I do, a passable arrangement in my brain. A kind of instinct that allows me to become aware of relations among data may well be connected with it."

Liebig first met Wöhler at Frankfurt in the winter of 1828. There they "ironed out" their previous difference of opinion over the

The Case of the Poisonous Socks: Tales from Chemistry
By William H. Brock
© William H. Brock, 2011
Published by the Royal Society of Chemistry, www.rsc.org

apparently identical composition of Wöhler's silver cyanate and Liebig's silver fulminate. They agreed (as Gay-Lussac had already suggested) that the two acids and their silver salts were remarkable examples of different modes of combination among the elements carbon, hydrogen, oxygen and nitrogen. In 1830, Berzelius coined the word *isomerism* to describe the remarkable phenomenon whereby organic compounds with very different chemical and physical properties were composed from the same elements in identical proportions but in some unknown different physical arrangement. In this respect, it is interesting to recall that Liebig had already puzzled over the polymorphism of calcium carbonate and argued that arragonite was identical with chalk. Berzelius, in defining isomerism, was mindful of an even more striking example that Wöhler had brought to light in 1828 when he showed that the product of reacting silver cyanate with ammonium chloride was urea (and *not* ammonium cyanate as he had expected).

In just ten years, between 1823 and 1832, we might say that Liebig and Wöhler between them transformed the nature and direction of organic chemistry when they were still both in their early twenties. Isomerism was to be, and remains today, a fundamental concept in understanding (and simplifying) the bewildering variety of the now millions of organic compounds that have been prepared by successive generations of chemists. Without this clarifying concept, organic chemistry would have long remained a "dark field" or "dark forest" as Liebig and Wöhler so aptly described it in 1832. Moreover, Wöhler's paper on urea of 1828 also did something else: it reinforced the possibility (already noted by a few other chemists like William Prout) that it should be possible to make organic compounds artificially—that is, as we would say today, it opened up the possibility of synthesis. Indeed, within thirty years, and during Liebig and Wöhler's lifetimes, it became possible for Kekulé and others to dissolve any difference between inorganic and organic chemistry and to redefine organic chemistry as simply the chemistry of carbon.

But let us return to the Frankfurt meeting between Liebig and Wöhler in 1828. Rather than being seen as youthful enemies, as their previous publications on cyanates and fulminates made them appear, the two chemists struck up a friendship that was to lead to the exchange of a remarkable correspondence starting 20 January 1829 and ending with Liebig's death in 1873. In these letters they exchanged details of work in progress, chemical and personal news, and gossip. They also agreed in Frankfurt (while Liebig was at Giessen and Wöhler was at the Berliner Gewerbeschule) to collaborate occasionally on research and to publish papers under their joint names—a series of nearly thirty papers that began with two papers in 1830, including one on silver cyanate. In his reminiscences Liebig recalled his good fortune in meeting Wöhler, for whom he developed the greatest affection. Whereas Liebig sought for similarities between substances, he noted, Wöhler always looked for differences. "The good that was in each became effective by cooperation" (Figure 17.1).

Collaborative research was extremely unusual at this date. We can recall, perhaps, individual joint papers by Lavoisier and Laplace, or Berzelius and Hisinger, or Gay-Lussac and Thenard; but these collaborations were almost always "one-off" arrangements and were never sustained in the manner of Liebig and Wöhler. Nor did Liebig collaborate extensively with anyone else: he published one paper with Gay-Lussac as an apprentice chemist in 1824; and a couple of papers each with Pelouze, Dumas, Mitscherlich, Pfaff, Redtenbacher and Will; but that was all. Although collaboration became increasingly commonplace in the nineteenth century and become ubiquitous in the big sciences of the twentieth century, sustained collaboration between the *same* two or more colleagues remains unusual even today. The only obvious examples that spring to mind are the research papers on physical organic chemistry of C. K. Ingold and E. Hughes between 1930 and 1960,

Figure 17.1   Friedrich Wöhler (1800–1882), professor of chemistry at the University of Göttingen and lifelong friend of Liebig's. (Wellcome Images)

and the collaboration between the Nobel Prize-winning bioche-mists S. Moore and W. H. Stein on amino acids. Thus, while Liebig and Wöhler's collaboration may have influenced the idea of joint publication, it has rarely been duplicated in a sustained way; most collaborations are between heads of research groups and successive waves of postgraduates. Liebig and Wöhler collaborated as equals for the joy of friendship.

If we were to adopt an incremental model of scientific progress, we would be forced to value equally every single publication by a scientist, and so accept that in the case of Liebig and Wöhler every individual or collaborative paper or book positively influenced the progress of chemistry, including industrial chemistry. Such an approach would be uncritical and absurd. Much of their work (as with the majority of chemists) was inevitably simply concerned with the first isolation, purification, definition and analysis of a substance which then became recorded in Gmelin's or Beilstein's *Handbuch der organischen Chemie*. Individual papers are, of course, sometimes important, and it would be foolish to deny that out of Wöhler's nearly 300 papers (roughly half of which were on organic chemistry) a goodly number were significant and influential. We might highlight those on the extraction of aluminium from alu-minium chloride (1828), the preparation of phosphorus from bones, sand and carbon (1829), the preparation of quinhydrone (1844) and of calcium carbide (1850), or the isolation of pure crystalline silicon and boron in 1857. The same would be true if selections were made from Liebig's 800 or so publications. His-torians of chemistry would argue that Liebig and Wöhler's joint papers on the benzoyl radical (1832) and on the transformations of urea (1838) were two of the most significant papers ever written in the history of chemistry.

The 1832 paper on the oil of bitter almonds (benzaldehyde) was outstanding and chemists the world over rightly perceived it as heralding a new dawn. It was a superb example of preparative chemistry, purification techniques, analysis, and the coordination of gravimetric results by the modelling of reaction schemes. The formulation of organic compounds as the chemistry of radicals—despite undergoing many changes at the hands of Dumas, Laurent and Gerhardt, and Kolbe and Kekulé—remains with us in the twenty-first century. The NMR and mass spectroscopy techniques now used in organic analysis rely essentially on the recognition of groups (or radicals) such as the benzoyl group defined by Liebig and Wöhler.

Similarly, the urea/uric acid paper of 1838 was of enormous significance chemically and medically, but also for Liebig personally since it brought him fame and recognition in Great Britain, led to his friendships with Faraday and Graham, and considerably altered the course of British chemistry. It is no accident that, four years after Liebig's first visit to the British Association for the Advancement of Science meeting at Liverpool in 1837 (where he read a preliminary version of the urea paper), Graham was enabled to found the Chemical Society in London in 1841. This magnificent investigation of the degradation products of uric acid had begun in June 1837 when Wöhler suggested that Liebig join him in investigating the chemistry of uric acid. Wöhler had just found that he had been able to extract what he thought was oxalic acid by heating uric acid with lead dioxide. Liebig repeated the experiment immediately and wrote back to Wöhler that it was not oxalic acid but allantoin.

It is interesting to note how allantoic acid was initially recognized by visual memory. Liebig had first prepared it from the allantoic fluid of a calf in 1829, ten years previously. Over the next few weeks they exchanged experimental data. Liebig wrote up the work in progress for *Poggendorff's Annalen* after Wöhler agreed that Liebig might also report their work to the British Association that September. The work, in which Heinrich Will assisted with nitrogen determinations, continued intermittently for another year. The huge 99-page paper, which identified and characterized sixteen new compounds as well as their many salts and derivatives, was written up by Wöhler who had gone to lengths to obtain pure uric acid from a boa constrictor. The paper was finally ready for publication in the *Annalen der Pharmacie* in June 1838, and concluded omnisciently:

> "The philosophy of chemistry will draw from this work the conclusion that that the production of all organic materials [*i.e.* materials found in living systems] in the laboratory, insofar as they no longer belong to the organism, must be regarded not only as probable, but also as certain. Sugar, salicin and morphine will be artificially produced. It is true that we do not yet know the way by which this conclusion will

be reached, since the precursors from which these materials develop are unknown to us, but we shall learn to know them in time."

The paper is particularly significant for historians since it looks forward to the approach taken to chemical physiology in Liebig's *Agricultural Chemistry* (1840) and his *Animal Chemistry* (1842).

Berzelius was ecstatic about the joint work, telling Liebig that it was one "of the most interesting and important contributions ever made to organic chemistry", while in his *Jahresbericht* for the year he noted the paper was as significant as their earlier paper on the oil of bitter almonds. "The wealth of new discoveries and substances analysed in the paper was without parallel".

While what I have called the cumulative approach remains significant to chemists and to the historian of chemical ideas and practices, it needs to be combined with a social approach. In this, we may view Liebig and Wöhler as at the centres of separate, but overlapping, national and international circles of communication and research: Liebig centred at Giessen and Wöhler at Göttingen. Both Liebig and Wöhler were highly influential teachers whose "state-of-the-art" laboratories at Giessen and in Göttingen attracted students from all over the world. Moreover, Liebig's development of a simple but rigorously accurate method of organic analysis in the early 1830s, and through his analytical text of 1837, and those texts of his pupils Will and Fresenius, together with Wöhler's *Übung in der analytischen Chemie* (1849), were enormously influential in directing the institutionalization and development of chemistry in Great Britain and America. We could say the same of Liebig's condenser, for although he did not invent this water-cooling distillation instrument, the promotion of its use in Giessen caused it to become (along with the five-bulb Kaliapparat), a ubiquitous piece of laboratory apparatus of the greatest value in preparative organic chemistry.

Liebig and Wöhler attracted huge numbers of students to both Giessen and Göttingen, and many of their pupils went on to create virtually identical laboratories and systems of teaching based upon competence in analytical chemistry in other countries. We think of the transposition of Giessen to Vienna, and above all to London with Hofmann's Royal College of Chemistry, and of Göttingen to Ira Remsen's Johns Hopkins University in the 1870s. Even if we

deny that Liebig and Wöhler invented the modern teaching
laboratory with its emphasis upon rigorous practical training, we
cannot deny their fundamental influence on the international
movement to build chemistry laboratories, to create research
schools of chemistry, and to promote the professionalization of
chemistry by insisting upon competence and qualification.

Liebig's decisive influence on the application of chemistry to
agriculture and to physiological chemistry in his two books of 1840
and 1842, and his outstanding success in popularizing chemistry
among a wider public through his *Chemische Briefe* beginning in
1843, is well known. It was not, of course, Wöhler's style to court
fame (or infamy) as Liebig did by publishing controversial books.
But we should note that apart from the good service he rendered
contemporaries by translating Berzelius's *Lehrbuch der Chemie* and
Jahresberichte into German, and by publishing his own influential
textbook of inorganic chemistry, Wöhler remained content at being
an "enabler" rather than a polemicist like Liebig. Nevertheless, in
his own gentler, quieter way, Wöhler was a significant "Wegweiser
moderne Chemie".

Unlike Liebig, who visited Britain seven times, Wöhler visited
England only once in 1835. Although his work was known to
British chemists either in its own right, or through his collaboration
with Liebig, or co-editorship of *Annalen der Chemie*, he never
cultivated British pupils or developed a controversial theoretical
programme as Liebig did. Moreover, because Liebig's teaching
methods were quickly adopted by the British in London and the
provinces, after about 1850 there was not the same need or excuse
for British chemists to study elementary chemistry in Germany.
Since American students still had this need, and because Liebig had
abandoned teaching on his move to Munich in 1852, they flocked
instead to Wöhler's laboratory at Göttingen. Wöhler himself estimated that he had taught something like 8,000 students during his
lifetime, at least 200 of whom were American.

Although Liebig's plan for *Annalen der Chemie* to be issued
simultaneously in Germany, France and England under the joint
editorships of Liebig, Wöhler, Dumas and Graham was quickly
abandoned (Wöhler saw it as a cheap "salesman's trick"), the
careers of Liebig and Wöhler and their pupils dramatically illus-
trate the growing internationality of science in the nineteenth
century. This is clearly seen in the way chemistry students travelled

from Britain or America to study in Germany, in the travels of Liebig and Wöhler themselves, their international correspondence, the dissemination of scientific journals and the diffusion of one country's scientific style to another. We can see this from Wöhler's textbooks. His text on inorganic chemistry of 1831 went through 15 editions. His equivalent text on organic chemistry (1840) went through thirteen editions, and both were translated into English and other languages.

Ironically, despite their having made fundamental contributions to the development of organic chemistry in the 1820s and 1830s, both Liebig and Wöhler turned aside from organic chemistry in the form that it was taking in the hands of their younger contemporaries. Wöhler turned increasingly towards inorganic and mineralogical chemistry and Liebig to applied chemistry. In Wöhler's case this was certainly because he was uninterested in chemical theory and the arguments about chemical modelling. In Liebig's case it is usually said that by 1840 he had become wearied by arguments with Dumas and Berzelius over theoretical interpretations of organic reactions and the paper chemistry formulae modelling of these reactions in terms of radicals, unitary type theories and electrochemistry, and that this is why he turned to something more domestic and human, namely the chemistry of agriculture and medicine.

In fact, Liebig's change of direction appears to have been caused by his dismay at the path organic chemistry was taking. Ironically, both Liebig and Wöhler were directly responsible for this direction. This direction was the creation of more and more *artificial* compounds that for them had no place or function in the organic nature of the plant and animal economy. In this interpretation, Liebig's turn towards agricultural and physiological chemistry in 1840 marked a determined return to *real* organic chemistry, *i.e.* the chemistry of organized, living or dead beings, or what today we call "natural products chemistry". Of course, the artificial organic chemistry (chemistry of things that do not exist in nature) set in motion by Liebig and Wöhler's outstanding papers on the benzoyl radical and the metamorphosis of uric acid (both actually written by Wöhler, according to Liebig) were soon to lead to the structural transformation of organic chemistry through the discovery of valence by Frankland in 1852 and by the development of structural theories by Kekulé and van't Hoff in the 1860s and 1870s.

The irony is that this same chemistry of artificial carbon compounds was to enable natural products chemists and biochemists
to have a far deeper understanding of living processes than was
possible for Liebig. In the twentieth century it came to transform
agricultural and physiological chemistry.

Nevertheless, how extraordinarily significant and important was
Liebig's commitment to natural products chemistry. He raised the
world's consciousness to the significance of the law of the minima,
and to the significance of mineral and man-made fertilizers in
maintaining and increasing food supplies. At the same time, he
inspired the medical profession to consider seriously the chemical
basis of health and disease. Between them, indeed, Liebig and
Wöhler had created a pathway through the dark province of
organic nature.

August Kekulé (1829–1896): Theoretical Chemist

In July 1867 the lively but short-lived weekly chemical journal *Laboratory* carried a specially commissioned essay from the pen of the 38-year-old German chemist August Friedrich Kekulé on one of the topics of the hour, the existence or non-existence of atoms. Kekulé's essay was a fairly ruthless critique of Benjamin Brodie's recently published "calculus of chemical operations" which denied Dalton's atomic theory and based chemistry entirely upon observable volumetric relationships using Greek letters as elementary symbols. The article clearly shows Kekulé as a powerful theoretician. Whether atoms really existed, he argued, was a metaphysical question: but atoms were conceptually useful in explaining chemical phenomena, and without doubt *chemical* atoms did exist. Besides explaining the constancy of chemical composition and Dalton's law of multiple proportions, the great strength of the atomic theory had been to explain why substances combined in the proportions observed and not others. For atomism had led to the idea of atomicity, the concept we now call valence. The article was to have had two more parts, but unfortunately *Laboratory* was aborted before they could be published. The historian is truly frustrated to learn that the third part was to have been concerned with graphic formulae.

The Case of the Poisonous Socks: Tales from Chemistry
By William H. Brock
© William H. Brock, 2011
Published by the Royal Society of Chemistry, www.rsc.org

The episode is interesting in what it reveals about the pulling power of the editor of *Laboratory*, John Cargill Brough, but also because it reminds us of the intellectual ties that bound Kekulé to Britain, and especially to the community of London chemists. It also reveals the extent to which almost all nineteenth-century chemists faced the possibility that elements and their constituent atoms were compound entities. In Kekulé's case such subatomic speculations may offer clues to the origins of the carbon chain concept for which he is best remembered.

FROM ARCHITECTURE TO CHEMISTRY

The Kekulés were an aristocratic Bohemian family whose Protestantism had led them to settle in the German states in the sixteenth century during the Thirty Years War. It was this Czech ancestry that was to justify Wilhelm II of Prussia's ennoblement of Kekulé in 1895, when he took the surname Kekule von Stradonitz. Note the dropping of the *accent aigu* or acute accent that had been used all his life to indicate to German speakers that the final *e* was pronounced (Figure 18.1).

Kekulé's family had long been resident in Darmstadt, the capital town of the Duchy of Hesse-Darmstadt. Here his father, Ludwig Karl (1773–1847), was a member of the ducal council and in charge

Figure 18.1 Friedrich August Kekulé (1829–1896), the distinguished German chemist whose adoption of the quadrivalence of carbon and a hexagonal ring formula for benzene ushered in the structural theory of chemistry. The portrait, by the Viennese artist Heinrich von Angeli, was commissioned by the German dye industry in 1890. Following his ennoblement in 1895 Kekulé took the surname of his Czech forebears, Kekule von Stradonitz. (Science & Technology Picture Library)

of munitions and war. Friedrich August (he rarely used his first name) was born on 7 September 1829, a child of his father's second marriage. A step-brother by an earlier marriage, Karl Kekulé, became a corn merchant in London. Kekulé received an extremely good education in the local Gymnasium alongside the sons of other court officials and evidently progressed better than Justus Liebig had done at the same school 25 years previously. Times had changed, and whereas Liebig's education had been entirely classical, Kekulé faced an impressively modern curriculum of classics, mathematics, science and modern languages, as well as art. A school-leaving report of 1847, published by Kekulé's biographer, Richard Anschütz, records "excellent, thorough diligence, knowledge and achievement" against his mathematics and natural philosophy, and commended his chemistry for "enthusiastic interest, commendable diligence and experimental dexterity". Kekulé left the Gymnasium at the age of eighteen in order to study architecture, his father's preferred career for him. He already had a good command of French, English and Italian as well as Latin, and his artistic skills were recognized as brilliant. One of his few surviving drawings, that of the courtyard of Heidelberg Castle, demonstrates his talent.

It was in June 1847 that Kekulé and his sister witnessed a fire in the Count of Görlitz's house opposite their home: it is often claimed that it was hearing Liebig demolish the defence case of spontaneous combustion of the Countess who perished in the fire, rather than of her murder by the family butler, that decided Kekulé upon matriculating at the University of Giessen, where another brother was already studying law. In fact, the case did not come to trial until 1850. Meanwhile he had already begun to study mathematics and undertake preliminary architectural studies at Giessen in the winter semester of 1847–1848, attending Liebig's lectures out of interest in the summer of 1848 (Kekulé's notes survive and have been published). However, with political upheavals imminent in the *vor März* period, and his father on his deathbed, Kekulé's family wanted him closer at hand and desired a more rapid route to his chosen profession. Accordingly, in the winter of 1848–1849 Kekulé began practical studies at the Gewerbeschule in Darmstadt (later the famous Hochschule). It was attendance at the analytical chemistry course of Friedrich Moldenhauer, a relative of Liebig's wife and one of the inventors of the safety match, that finally persuaded Kekulé to abandon architecture for chemistry—something that was easier to do following his father's death. Accordingly, in the summer of 1849 Kekulé joined Liebig's laboratory at Giessen, being principally trained by the assistants Theodor Fleitmann, Adolphe Strecker and Heinrich Will. Liebig was then actively engaged in revising the third edition of his famous *Chemical Letters* and, as a result, Kekulé was commissioned to make ash analyses of gluten and wheat bran, which Liebig published as appendices to his book in 1851. These formed Kekulé's first publication. In addition, Kekulé made an investigation of amylsulfuric acid and its salts for Will which gave him another publication and the award of a doctorate in 1852.

Historians are so used to stressing the way British chemists flocked to Germany in the nineteenth century that they tend to ignore the fact that before the take-off of large-scale chemical industry, the German states were over-producing chemists. With no immediate prospect of an industrial or academic position, it was Liebig who advised Kekulé to undertake a period of further study in Paris—just as he himself had done in 1819. Kekulé spent the period May 1851 to April 1852 in France armed with introductions to Liebig's French pupils Regnault and Pelouze, and advice to study gas analysis and assaying techniques with them. At the beginning of his stay, however, Kekulé came across Charles Gerhardt's *Introduction à l'Etude de Chimie par la Système Unitaire* (1848) and was sufficiently impressed by the book to attend a series of lectures that the author was giving. Gerhardt, who noticed Kekulé's name among his subscribers, quickly identified his auditor as the analyst quoted by Liebig in his recent *Chemical Letters*. In consequence they became firm friends and Kekulé adopted the type theory, a method of classifying or representing molecules that required the assumption that a molecule of water contained two atoms of hydrogen and one of oxygen. Such standardization brought order to the hitherto seemingly incompatible molecular magnitudes deployed in inorganic and organic chemistry. Equally important for Kekulé's intellectual development was his friendship with another disciple of Gerhardt's, Adolph Wurtz.

THE LONDON EXPERIENCE

In April 1852 Liebig informed Kekulé that his wealthy former pupil, Adolf von Planta, needed an assistant to help in his private laboratory in his castle at Reicheau bei Chur in Switzerland. Von

Planta was interested in plant alkaloids and got Kekulé to investigate nicotine and conine, as well as local mineral waters. All this was pretty routine stuff, and although Kekulé was never an outstanding practical chemist, he served his apprenticeship well and had plenty of time to think about French ideas of classifying organic compounds. Further correspondence with Liebig at this time shows Kekulé desperately hoping his former teacher would invite him to assist at Munich (where Liebig had moved that year); instead Liebig found Kekulé a post with John Stenhouse in London at St Bartholomew's Hospital. Stenhouse, an English pupil of Liebig's, was partly paralysed and needed help with preparations and demonstrations in his lectures to medical students. Reluctantly, for he had no enthusiasm for the role of "grease chemist" (*Schmierchemiker*), Kekulé realized that an English experience would be useful. There was a step-brother in London and a German chemical community that included August Wilhelm Hofmann, Hugo Müller and another of Liebig's recent pupils, Heinrich Buff. In the event, because of his French connections, he rapidly made friends with Alexander Williamson at University College (yet another of Liebig's pupils) and William Odling, a fellow hospital chemist who was busy translating Laurent's *Chemical Method* into English. Both Williamson and Odling were Gerhardtian *typists*. In its more mature phase this classificatory scheme recognized that elements fell into four main categories: monobasic (or monatomic), di-, tri- and quadri-basic (for example, H, O, N and C) and from these, four *types* of compound were derived, *i.e.* HH, OH_2, NH_3 and CH_4. Furthermore, from these basic types, chemists could derive subtypes such as HCl or SH_2, as well as multiple or mixed types where a polybasic atom provided a bridge by replacing two or more atoms of hydrogen or its monobasic equivalent.

At Williamson's suggestion, and with Stenhouse's approval, Kekulé began to investigate the reaction of phosphorus pentasulfide on acetic acid, a member of the water type. The product, thioacetic acid (the first known organic thio acid) could either be classified as a water type or, as suggested by Kekulé in 1854, as a member of the derived hydrogen sulfide type. It was from such classificatory games that the notion of substitution emerged as well as that of elements and radicals having equivalent values of substitution or "equivalence". From this it was but a small step to the concept of "valence" or "valency", the belief that each atom had a definite combining power.

FROM TYPES TO CATENATION

With Kekulé still no closer to a university chair, Liebig and Bunsen recommended that he should settle in a university town. Kekulé chose Heidelberg to be close to Bunsen and where he could offer private lectures to students as a *Privatdocent*. Kekulé was able to set up a laboratory in his apartment and to employ as an assistant Adolf Baeyer, who was to become as famous as his employer. In Heidelberg, Kekulé and Baeyer investigated the composition of mercury fulminate, using Odling's concept of a marsh gas, or methane type, CH_4. (Kekulé's formulae of 1867 are modernized for clarity.)

CHHHH	CHHHCL
marsh gas	methyl chloride
CHClClCl	$C(NO_2)HgHg(C_2N)$
chloroform	methyl fulminate

More importantly, Kekulé further refined an idea that had occurred to him in London, that of polyvalent radicals.

In his great paper of 1858 "On the constitution and metamorphoses of carbon compounds and on the chemical nature of carbon compounds", in which he extended the equivalence of carbon to all its compounds, Kekulé stressed how indebted he was to the English and French schools of chemistry for his interpretation. With hindsight we can see that at one stroke organic chemistry had been unified. Chemists no longer needed separate types for paraffins, ethers, amines, *etc.*; all organic compounds were now embraced within the idea of carbon chains (catenation) and the notion of carbon's tetravalence. How had this insight occurred to Kekulé?

The American historian of chemistry, Alan Rocke, has suggested that the concept of structure (the application of valence rules to the presumed construction of molecules) probably owed much to Wurtz's work. In his influential book, *Chemical Method* (1853), August Laurent had speculated that atoms might be divisible in order to explain why, for example, iron had both odd and even powers of combination to form both ferrous and ferric salts. At the same time, in England, Alexander Williamson and William Odling were developing the idea of double, triple and mixed types in which one dibasic (diatomic) molecule linked together two monobasic radicals. For example, in sulfuric acid, the SO_2 group could be said to link two hydrogen atoms together, having substituted or exchanged itself for the one oxygen atom within the water type. In 1855, following Laurent's hint, Wurtz wondered whether oxygen was dibasic and nitrogen tribasic because these elements were formed from two or three juxtaposed sub-atoms. According to this conception, polyvalent atoms were really aggregates of monovalent sub-atoms. As Rocke has pointed out, this is much the same as the later school textbook rule that an equivalent is "atomic weight divided by valence", and it is still reflected in the definition of atomicity as the number of atoms in a molecule (rather than of sub-atoms in an atom).

So, when Kekulé visualized dancing atoms in his reverie on a London omnibus in 1855, did he imagine segmented worm-like entities made up of sub-atoms? The "sausage-shaped" graphic formulae that he first used in lectures at Heidelberg in 1857 certainly suggest so and seems to be confirmed by Kekulé's statement in 1867 that "polyvalent atoms, with respect to their chemical worth (valence), can be viewed in a sense as a conglomeration of several monovalent atoms". Since Wurtz's paper appeared in the summer of 1855, Rocke conjectures that Kekulé's image of linked carbon atoms most likely occurred on the London omnibus in the late summer after he had read and reflected upon Wurtz's paper. Kekulé's graphic formulae, which he used in his textbook, were visualizations of Wurtz's speculation. Oddly, Kekulé rarely used these sausage formulae in his published papers, where he continued to use type formulae as a means of classification.

For Kekulé, type formulae merely compared "various compounds with one another relative to their chemical composition". They were not illustrative of the real constitution of compounds. This is clear from the way the "typed" isomers ethyl hydride, C_2H_5} and dimethyl, CH_3} are dissolved when they are rendered as a

$$\left. \begin{array}{c} C_2H_5 \\ H \end{array} \right\} \qquad \left. \begin{array}{c} CH_3 \\ CH_3 \end{array} \right\}$$

carbon chain $H_3C–CH_3$ provided that all four valences of carbon are identical, as was proved by Carl Schlorlemmer in 1864.

It was Alexander Crum Brown, and especially Edward Frankland, who slowly weaned the chemical community towards our familiar graphic formulae in which the "carbon chain" property that was implicit in Kekulé's concept of quadrivalent carbon was made explicit. Frankland's formulae, apart from being simpler to print, possesses the great advantage of being able to represent the three-dimensional character of molecules. The heuristic character of Kekulé's formulae may have been responsible for his refusal to accept that valence could vary. Although proved wrong, nevertheless it was his conviction that carbon's valence was invariable that logically led him to posit the carbon chain, benzene rings, and double and triple bonding.

THE GOOD NEWS FROM GHENT

To provide a critical focus for this new constitutional or structural approach to chemistry, Kekulé and other *Privatdocenten* at Heidelberg launched the periodical *Kritische Zeitschrift für Chemie* in 1858. Kekulé soon ceded the editorship to Emil Erlenmeyer in whose hands it became a major journal and an important vehicle for Russian chemists to communicate with their European colleagues.

Kekulé's paper on the carbon chain made him famous throughout Europe, with the result that in 1858 he was called to the University of Ghent in French-speaking Belgium, where Jean Servais Stas, of atomic weights fame, was striving to strengthen science teaching in his country. Kekulé's seven *Wanderjahre* were over and he was now at the peak of his productivity as a chemist. In Ghent he began to write his influential, but incomplete, textbook, *Lehrbuch der organischen Chemie* (fascicle 1 appeared in June 1859; the first and only installment of the fourth volume appeared in 1887). Here Kekulé grouped compounds into three broad classes of aliphatic, and benzene and naphthalene derivatives based upon general formulae such as C_nH_{2n}, *etc.* As his British pupil Francis Japp said, "The effect produced by the book was enormous. The facts of organic chemistry began to group themselves spontaneously under the new system". This was the system of classification adopted by Beilstein for his *Handbuch der organischen Chemie* which continues to be the bible of organic chemists.

The "quiet revolution" in atomic weight standardization based upon chemists using "two-volume formulae" (that is, agreeing that the formula of water was H_2O) needed consolidation and international agreement. This was Kekulé's purpose in initiating an international congress of chemists, which he persuaded his friend Karl Weltzien to organize at Karlsruhe in September 1858. It may be recalled that it was at this meeting that Stanislao Cannizzaro urged chemists to base their quantitative reasoning concerning atoms and molecules on Avogadro's hypothesis of 1811 that, under the same conditions of temperature and pressure, equal volumes of gases contain the same number of molecules.

At Ghent, Stas and Kekulé collaborated on the design of spacious new laboratories which were opened in 1861. Using these facilities Kekulé began to study the isomers, fumaric and maleic acids, and their bromine derivatives which (in 1862) he interpreted in terms of double (or unsaturated) bonds between adjacent carbon atoms in the former and free affinities in the latter. (The geometrical nature of their isomerism was not explained until 1874, when J. J. van't Hoff solved the problem.) The concept of double bonds proved useful in the next few years as Kekulé struggled to understand and to classify aromatic compounds for his textbook. No one can be exactly sure when Kekulé hit on the closed chain of six carbon atoms linked by alternating single and double bonds for the structure of benzene; but it probably occurred to him at the same time of his marriage in 1862. At the end of his life he told the story of dreaming that, like a snake eating its own tail, he had imagined a chain of dancing carbon atoms forming a closed circle. There has been much interest and controversy over Kekulé's claims for this inspiration. Given the subconscious wellsprings of many scientific ideas it is not inherently implausible, even if it was related in the rhetorical context of an after-dinner speech 30 years afterwards.

The idea of unsaturation was controversial enough, without the additional suggestion that unsaturated aromatics like benzene formed a common hexagonal nucleus in many such compounds. In 1864, therefore, Kekulé hired two assistants, Karl Glaser and Wilhelm Körner, both Giessen-trained chemists, to help him test the hypothesis. Only in January 1865 did Kekulé feel confident enough of the hexagonal benzene sausage formula which was first announced to the Paris Chemical Society. The more familiar hexagon followed a few months later. By then Kekulé knew that Hofmann, who had been called from the Royal College of Chemistry in London to the University of Bonn, had decided to return to teach at the University of Berlin instead. Kekulé let it be known that he was keen to return to Germany and to teach in his native language again. The Prussian government duly appointed Kekulé to the chair at Bonn in September 1867.

ANTICLIMAX IN BONN

The final phase of Kekulé's career, from 1867 until his death nearly 30 years later, was something of an anticlimax, as many chemists commented in their private correspondence. Never as prolific as his older or younger contemporaries such as Liebig, Bunsen, Hofmann, Kolbe or Baeyer, there was nevertheless a distinctive flagging of energy once he had moved to Bonn. By 1874, despite the splendid laboratory facilities that Hofmann had designed for himself but never used, and the presence of able research students such as J. H. van't Hoff from the Netherlands, T. E. Thorpe from England and Theodore Zincke, Hermann Wichelhaus and Richard Anschütz from Germany, productivity was very much lower than from the "steam intellect" chemistry institutes of Munich, Berlin, Leipzig and Heidelberg. Production of the textbook was unremittingly slow and from 1874 onwards Kekulé scarcely managed a paper or a published speech a year. Of course, Kekulé was not alone in demonstrating waning powers; in England, Williamson published nothing after about 1864, not dying until 1904, while Odling, who assumed the Waynflete Chair of Chemistry at Oxford in 1880, published nothing chemical at all between then and his death in 1921.

Despite his lethargy, which Kolbe used in inexorable criticism of Kekulé's structure theory, Kekulé did in fact make some significant achievements at Bonn, including in 1872 the dynamic oscillation formula for benzene to explain the lack of isomeric disubstituted derivatives that otherwise seemed permissible. In the early 1880s, when several alternative possibilities had been touted for the structure of benzene, including the prism formula proposed by August Ladenburg, Kekulé demonstrated something of a return to his old powers when he showed that a series of experimental metamorphoses of pyrocatechol and quinine into tetraoxysuccinic acid and β-trichloacetylacrylic acid were explained more simply if benzene was assumed to be hexagonal. Again, in 1890, he brilliantly summarized the chemistry of pyridine, which his pupils

Körner and James Dewar had shown to be a benzene ring with nitrogen substituted for one CH group; while in 1892, four years before his death in 1896, he was the first person to prepare absolutely pure formic aldehyde.

FAMILY LIFE AND DEATH

In June 1862 Kekulé married Stephanie Drory, the daughter of an English gas engineer who was employed in Belgium; but his wife died in May 1863 during the birth of their son, Stephan. Kekulé remarried in October 1876 to Luise Högel (1845–1920) who had been his housekeeper in Bonn. There were three more children of this marriage but, whether because of his premature ageing and deafness, or his wife's irascibility, by all accounts the relationship proved an unhappy one. Kekulé died on 13 July 1896 following an attack of influenza and was buried in the family vault in the Poppelsdorf cemetery close by the university. A bronze statue commissioned by the German dyestuffs companies was erected outside the Bonn Chemical Institute in 1903. Later, in 1927, the indefatigable efforts of his devoted pupil and biographer, Richard Anschütz, led to the opening of a Kekulé museum and library in the Darmstadt Hochschule (now the Technische Universität Darmstadt).

The historian of valence, Colin Russell, long ago pointed out that although it is convenient to suppose that it was Kekulé's architectural training that helped him to conceive molecular structures, what is more striking is the view he acquired from Williamson of the dynamic nature of molecules. Architecture is essentially static, whereas Kekulé's conception of structure was much more fluid and imprecise. As Russell concluded, "It seems better to depict Kekulé as the restless visionary contemplating the giddy dance, the writhing serpents and all the ceaseless activity which symbolizes for him, and us, the world of atoms".

CHAPTER 19

The Don Quixote of Chemistry: Sir Benjamin Collins Brodie (1817–1880)

The English chemist Benjamin Collins Brodie, who was regarded by Kekulé as "definitely one of the most philosophical minds in chemistry", was the eldest son of Britain's leading physiologist and surgeon, Sir Benjamin Collins Brodie (1783–1862). Brodie *père*, who was president of the Royal Society from 1858 to 1861, had been made a baronet in 1834 for his medical services to the Royal family and his son inherited the baronetcy in 1862. A theist and anti-materialist, Brodie senior was profoundly interested in metaphysical questions. He published two volumes of *Psychological Enquiries* (1854 and 1862), a series of dialogues between a country gentleman, a doctor and a lawyer, that were much influenced by Sir Humphry Davy's posthumous *Consolations in Travel* (1830). These well-meaning, but ultimately turgid, dialogues were concerned with unfashionable topics such as mind–matter dualism, natural theology, and the problems of pain and immortality. They seem to have made little impact on Brodie's contemporaries, who were finding Herbert Spencer's psychological and evolutionary writings more exciting. However, their publication suggests that the younger Brodie was brought up in an atmosphere of philosophical inquiry in which the metaphysical foundations of scientific beliefs were critically questioned.

The younger Brodie was educated at Harrow School from where he won a classics scholarship to Caius College, Cambridge. However, his father, preferring him to be educated as a commoner, as befitted the family's status, sent him to Balliol College, Oxford, in 1835. There, under the influence of the mathematical physicist Baden Powell, his interests turned away from the classics to mathematics. He also attended the chemistry lectures given by Charles Daubeny in the basement of the Ashmolean building opposite Balliol (now Oxford's Museum of History of Science). Brodie graduated in 1838, but because of his refusal to assent to the 39 Articles of the established Church of England, he was unable until 1860 to obtain the MA degree essential for a respectable academic career at Oxford and he was always denied a college fellowship. For some time after graduation Brodie trained for the bar at Lincoln's Inn in the chambers of an uncle. In 1844, however, he met Justus von Liebig as a guest in his father's house and immediately abandoned the law to study chemistry at Giessen, where he was awarded a doctorate several years later in 1850 for the analysis of beeswax. This work, for which he also gained the fellowship of the Royal Society as well as its Royal Medal in the same year, proved the existence of solid alcohols that were homologous with known alcohols, and had important implications for the understanding of animal metabolism (Figure 19.1).

In the decade following his return to England in 1845, Brodie worked in his own private laboratory in Albert Road, near Regent's Park in London, where he taught chemistry to his friend and later Oxford mineralogical colleague, Nevil Story Maskelyne. In 1847 he joined the Royal Institution as an assistant to William Brande (a close friend of his father's) where he came into contact with Michael Faraday whose negative views on atomism probably influenced him. In 1844 Faraday had rejected atomism because of the conundrum why, if it contains more atoms per unit volume, is potassium hydroxide a non-conductor, whereas potassium, with fewer atoms, is a conductor? On Brande's retirement in 1853 Brodie hoped to succeed him and to transform the Royal Institution into a research institution on the Liebig–Giessen model, but he was strongly opposed by the managers who disapproved of the "advanced" and unpopular character of his lectures. They were too difficult for popular consumption.

Figure 19.1   Sir Benjamin Collins Brodie (1817–1880), the son of a famous
surgeon of the same name. Brodie studied with Liebig in Germany
before becoming professor of chemistry at Oxford. A highly
competent chemist best known for his ozonizer, he became
notorious for his mathematical alternative to atomism, the calculus of chemical operations. (RSC Library)

   By 1850 Brodie had established himself as a leading experimental
and theoretical chemist. Historians of chemistry have classed him
as one of the circle of modernizers who accomplished the "quiet
revolution" of systematically basing the construction of both
inorganic and organic molecules on two volumes of hydrogen, a
view principally propelled in London by Alexander Williamson at
University College and August Hofmann at the Royal College of
Chemistry during the 1850s. Brodie's early chemical work, which
was implicitly atomistic, was an attempt to reconcile Berzelian
electrochemical dualism with the most recent Gerhardtian
ideas concerning the self-combination of atoms (*i.e.* two-volume
formulae for hydrogen, H_2, chlorine, Cl_2, *etc.*). He was intensely
interested in allotropy, which he believed to be due to the
arrangement and electric charges of particles making up an element. He discovered that iodine catalyzed the conversion of yellow
into red phosphorus, and that pure graphite, when treated with

potassium chlorate, formed a crystalline graphitic acid which he speculated might contain a graphite radical (Gr_4), or *graphon*. His process for the purification of graphite, which he patented, proved of considerable technical value. He was secretary of the Chemical Society from 1850 to 1856, and its president from 1859 to 1861. He was one of the British delegates to the conference on molecular weights in Karlsruhe.

In 1855, despite considerable opposition from theological fellows, Brodie succeeded Daubeny as Professor of Chemistry at Oxford, where he did much to gain recognition for chemistry as an academic study, as well as proper laboratory facilities for its teaching. He had long been friendly with an influential group of Broad Church clergyman dons such as Benjamin Jowett and Arthur Stanley who, however, disapproved of his atheistic tendencies. Another Oxford influence was Richard Congreve who (like Williamson) had studied with Auguste Comte in Paris and espoused the cause of Positivism in Britain. Exposure to positivist thought in Oxford, together with his atheistic tendencies regarding revealed religion, Faraday's dismissal of atomistic explanations, and Laurent's and Gerhardt's espousal of the unity of chemical theory based on rational and systematic language, all seem to have caused Brodie to conclude that atomism was leading chemists astray.

At the beginning of the 1860s Brodie turned his back on the structuralist tendency of organic chemists such as Williamson and Adolph Wurtz and professed a determined scepticism towards the truth and conventional utility of the atomic theory. His sustained opposition to Dalton's atomism during the last twenty years of his life proved the most remarkable philosophical and theoretical achievement of his career. As a positivist dedicated to the removal of the metaphysical from science he strongly objected to the realism implied by the availability of molecular models made of balls and wires that contemporary instrument makers had placed on the market following the lecture demonstrations of Hofmann, and the bonding symbolism introduced by Alexander Crum Brown and Edward Frankland.

Brodie's position was that the ultimate nature of matter was unknowable; chemistry had to be based solely on observable phenomena. For Brodie, influenced by his reading of Lavoisier, Condillac, Gerhardt and Comte, atoms were unnecessary and a confusing interpolation between observation and expression of

phenomena because they were not subject to any rules and invited
the unwary to think of chemical phenomena in terms of real balls.
He denied that the object of science was to explain. He agreed with
Gerhardt that "chemical formulae are not meant to represent the
arrangements of atoms, but rather to make evident simply and
exactly the relations that link bodies during transformations". We
cannot ask what water is, but only describe how it behaves and what
it becomes after interaction with other chemical materials. Because
we have no way of grasping the underlying reality of things, we must
be content to describe accurately how matter behaves.

Because Daltonian–Berzelian atomism had led chemists astray,
atomism and its symbolism had to be swept away. The facts of
chemistry were to be represented by suitable symbols that could be
derived algebraically from Gay-Lussac's law of volumes and
Dulong and Petit's law of specific heats. Further algebraic
manipulation of the symbols might then lead to new insights.
"Such a system", he claimed, "is based, in the most absolute sense,
upon fact, for it presents only two objects to our consideration, the
symbol and the thing signified by the symbol, the object of thought
and the object of sense".

In 1866 the Royal Society began to publish Brodie's *Calculus of
Chemical Operations* which introduced Greek letter symbols for the
chemical elements to replace the Roman alphabet (Berzelian)
symbols that contemporary chemists used to represent atomic
weights. Brodie's symbols, however, represented *operations* on
space (*volumes*) not weights for, besides its revolutionary symbolism, the calculus also demanded an appreciation of George Boole's
algebraic logic, which Brodie had studied after the publication of
Boole's *Investigation of the Laws of Thought* in 1854. In this, an
equation such as $y = xy$ is a symbolic statement that y is a subset of
x in which the symbol x is an operator on y. Although professional
mathematicians like Oxford's William Donkin and Henry Smith
later advised Brodie, it appears that he developed the system
without professional help. The principal difficulty about the
calculus for the present-day historian and philosopher of science is
the need to explain it before going on to discuss it and the difficulty
of giving any concise description of it.

Boole had developed the concept of symbolic operators in
algebraic analysis. These provided a code as to how symbols were
to be understood and manipulated. Brodie exploited this in the idea

of a chemical operator, or chemical operations, that he symbolized by Greek letters. Brodie proposed that if two substances with the empirically derived weights, x and y, combined to form a new compound with weight xy, then $x + y = xy$. From such weight equations he constructed a symbolic algebra that bypassed any atomistic interpretation. Aware from recent chemical history that an absolute standard of comparison was required, he chose for volumes the "litre", which he defined as a unit of space (analogous to Boole's universal set). A choice of standard element was also required, and like Dalton, he chose hydrogen. However, he defined the element (α) as having a simple weight of one—in other words, he did not allow it to be distributed in chemical operations (reactions). In molecular terminology, his standard was $H = 1$, and not $H_2 = 1$. The assumption meant that all elements of odd valences had to be symbolized by a combination of prime factors, one of which was α. Thus chlorine became $\alpha\chi^2$, *etc.* For nonvaporizable elements, Brodie made use of Dulong and Petit's rule, together with additional assumptions. The resultant system generated three kinds of elementary symbol:

(1) Those like hydrogen and mercury expressed by a single symbol (*e.g.* α).
(2) Those like oxygen and sulfur expressed by identical symbols (*e.g.* ξ^2).
(3) Elements such as the halogens that appeared to be a combination of the first two groups (*e.g.* $\alpha\chi^2$).

Brodie justified the simple assumption that hydrogen was undistributed by arguing that it predicted the law of even numbers, whereas an assumption that hydrogen was α^2 did not (though it was compatible with the law). Contemporary chemists were quick to point out that if hydrogen was allowed a compound weight α^2, then all the Greek symbols would become formally identical to those of Berzelian atomism (*viz.* $\alpha^2 = H_2$). In this light, the question of the Proutian complexity of the elements became something of an *experimentum crucis* for the calculus.

Few contemporary chemists were able to follow Brodie's mathematical reasoning and what principally interested them was its implication that elements like chlorine might be compounds that *contained* hydrogen. The new spectroscope appeared, at first, to promise validation of Brodie's system when viewed as predictions. His "ideal chemistry", as he called it, stimulated a good deal of fruitful controversy in the 1860s and 1870s, but it ultimately foundered because of Brodie's inability to account for the phenomena of structural isomerism and stereoisomerism. Both properties and methods of preparation distinguish isomers; but in the calculus methods of preparation are unimportant, as long as the same compound results. Although Brodie struggled with probability theory, his notation refused to yield a simple method of distinguishing isomers—something that was brilliantly elucidated by Le Bel and van't Hoff using the model of chemical structure based upon an atomic theory of matter that carbon was quadrivalent. Nor, as it transpired, did Brodie's three groups of elements (differentiated by the form of their symbolism) bear any analogy to the groupings produced by the periodic law.

Brodie resigned from Oxford in 1872 because of ill-health and retired to a magnificent house on the top of Box Hill in Surrey. In the same year he published a paper on the action of electricity on oxygen which confirmed the calculus suggestions that the ozone molecule was triatomic and introduced the well-known apparatus for the preparation of ozone, "Brodie's ozonizer". He died at Torquay on 24 November 1880 from rheumatic fever, with the calculus on which he had spent twenty years of his life uncompleted.

One historian has seen Brodie as the Don Quixote of chemistry, tilting his mathematical lance against the windmill of atomism. Although a chemical *cul de sac*, Brodie's calculus of operations nevertheless remains of interest to historians and philosophers of chemistry. In the first place, his methodological use of the thought experiment is interesting. Thus, in seeking support for the possible existence of unknown primitive elements such as χ (*chi*) (which spectroscopy might reveal as present in the Sun), he imagined (like Swift) a country called Laputa where carbon could not be isolated because experiments could not be conducted between $0\,^{\circ}\text{C}$ and $300\,^{\circ}\text{C}$. Yet, in using the calculus of operations, Laputian chemists might speculate that carbon existed from the derivation of the symbol $\alpha\kappa^2$ (alpha kappa squared) from their experimental ability to reduce two units of methane to three units of hydrogen and one of acetylene. Brodie was keen to use thought experiments to support the compound nature of chlorine, $\alpha\chi^2$. He became very excited in 1879 when Victor Meyer thought he had detected the presence of oxygen in chlorine—a discovery he soon retracted, to Brodie's disappointment.

Secondly, mathematically speaking, Brodie used a technique called normalization. In order to classify and factorize chemical equations (the burden of the second part of the calculus published in 1877), Brodie had to "normalize" his equations with respect to space. He did this simply by adding a numerical factor (representing units of empty space or null sets). Such mathematical manipulations of equations were not to reappear in chemistry again until the advent of quantum chemistry in the 1930s. Brodie can be seen to be a pioneer in believing the possibility of finding mathematical solutions to chemical problems.

Thirdly, his calculus implicitly involved him in assessing possible mechanisms of reactions. Although fellow chemists and type theorists such as Williamson had begun to study how organic reactions worked, Brodie seems to have been the first to state explicitly how his symbolism helped explain the likely mechanism of an operation. Given the relative simplicity of Brodie's Greek symbolism, it was easy to "see" the shifts and substitutions that were taking place. (In fact, of course, it is also possible to factorize molecular equations which, in effect, were what the Crum Brown–Frankland structural formulae notation made visible.)

Finally, Brodie's calculus foreshadows recent insights into the historical significance of chemists' ability to manipulate symbols on paper or in today's computers. Brodie identified chemical equations as "a study of transcendental interest" insofar as they yield new truths. It is only recently that what has been called "paper chemistry", and the viewing of chemical formulae as tools and instruments, have received attention by chemical philosophers.

CHAPTER 20

The Epistle of Henry the Chemist

The organic chemist Henry Armstrong was a regular contributor to the pages of *Chemistry & Industry* in the early part of the twentieth century. In May 1932 its readers were faced by a typical tirade.

> "Whilst the academically minded unadventurous Fellows of the Chemical Society are comatose—lost in dreams of tadpole formulae, leaning protons, electron sinks, and other frippery for which the world cares not a jot—without any proper educational spirit and scarce an original idea, let alone any sense of proportion or of practical value; and whilst too the Royal Society equally high-bow'd and withdrawn from the world, is developing the new exciting game of atomic skittles (let us hope it may not prove to be 'all beer' also): the Royal Society of Arts, the one practical body left to us, mindful of great men like Prince Albert and Lord Playfair who ever had the public welfare in mind, true to its traditions, seizes on an all-chemical topic, that of *vitamins*—the most important topic that could possibly be considered—for its spring-cleaning course of Cantor lectures, choosing as exponent the man who better than anyone, at the moment, is able to make a clean sweep of the board—Professor Drummond."

The paragraph is vintage Armstrong. Penned at the age of 84 it snarls at many of his twentieth-century bogies: ions and electronic

The Case of the Poisonous Socks: Tales from Chemistry
By William H. Brock
© William H. Brock, 2011
Published by the Royal Society of Chemistry, www.rsc.org

theory in chemistry, nuclear bombardment in physics, scientific jargon, the loading of education towards theory instead of laboratory and workshop practice, and the fact that scientific societies had ceased to concern themselves with really pressing problems, like food and coal.

Earlier, in 1929, he had dismissed a paper of C. K. Ingold's as "jargonthropos". Quoting Ingold's long title: "Influences of poles and polar linkings on tautomerism in the simple three carbon system. Part I. Experiments illustrating prototropy and anionotropy in trialkylpropenylammonium derivatives", he had continued:

> "Every word in this strange cacophonous medley needs explanation ... The theme is trivial—merely the production from the compound $CH_2Cl.CH.CH_2Cl$ of the two simple chloroallylammonium derivatives and the action of sodium ethoxide thereon. The behaviour brought out [namely] the shift of the ethenoid [double bond] junction is that well known to be characteristic of such compounds and is in no way peculiar or momentous. The work at most adds a bare item or two to Beilstein. The discussion is superficial—a wordy paraphrase of the few simple statements which would have been held to be significant in the 1870s and 1880s."

As for Robert Robinson's and Ingold's electronic arrows, they were dismissed with the brilliant witticism that "bent arrows never hit their marks".

THE CHEMICAL CRITIC

Historians love to quote Henry Armstrong—the man who got it wrong about the rare gases by insisting that Ramsay had mistaken argon for an allotrope of nitrogen; the chemist who (while not entirely wrong) set his heart against the three musketeers of Arrhenius, Ostwald and van't Hoff and their ionic theory. As a journalist would say, Armstrong is good for copy. He is only too easily dismissed, like his teacher Hermann Kolbe, as an old fogy and stick-in-the-mud whose knowledge of chemistry and physics got stuck in a time warp about 1894. But just as Kolbe has been brilliantly rehabilitated and explained by Alan Rocke in a superlative biography, perhaps it is time to do something similar for Henry Edward Armstrong.

In fact, the comparison between Kolbe and Armstrong is unfair and inept, since Armstrong was never cold-shouldered by his fellow chemists or felt to be an embarrassment in the way that Kolbe was. Instead they welcomed Armstrong's comments at scientific meetings and were delighted to see his highly literate and scintillating letters and essays in the pages of *The Times, Chemical News, Nature* and *Chemistry & Industry* during his long retirement from 1912 to 1937. His last paper, the year he died at the age of 89, was on the value of weeds—a reminder of the strong interest he took in agriculture and horticulture from the time of his election as a Trustee of the Lawes Agricultural Trust at Rothamsted in 1889.

TRAINING

A Falstaffian bewhiskered figure, this genial English chemist who was born at Lewisham in 1848 was a pupil of August Kekulé's two contemporaries and rivals, Edward Frankland and Hermann Kolbe. Armstrong ranked Frankland, with whom he studied at the Royal College of Chemistry in Oxford Street from 1865 to 1867 and after whom he christened his first son Edward Frankland Armstrong, alongside Joseph Lister and Louis Pasteur as "a saviour of the world's health"; and Kolbe with whom he studied in Leipzig between 1867 and 1870 as "the real parent of the modern system of structure and formulae". Elsewhere, he reminisced that he and Horace Brown (later a brewery chemist) were the first of Frankland's students at the Royal College of Chemistry after he had succeeded A. W. Hofmann, and they had been used by him as immediate tests of the new graphic system of formulae which Frankland published shortly afterwards in his *Lecture Notes for Chemical Students* (1866). "We became past masters in the art of constructing 'imaginary' constitutional formulae. Consequently, structural chemistry grew into a habit" (Figure 20.1).

Since, as a student, Armstrong also attended the lectures of T. H. Huxley and John Tyndall at the School of Mines in Jermyn Street, we should note that he was associated with all three of the colleges that eventually constituted Imperial College in 1907. Today, Armstrong (who saw himself as born an Englishman but "made in Germany") is more remembered for his campaigns to improve the

Figure 20.1   Henry Edward Armstrong (1848–1937), a British organic chemist
trained by both Edward Frankland and Hermann Kolbe. (RSC
Library)

teaching of chemistry and science by the deployment of the heuristic or discovery method of instruction, and less for his theory of
residual affinity or, as I have indicated, for his hostility towards
physical chemistry and, in particular, the theory of dissociation.
However, right up to the time of his retirement in 1912 from the
Central College of the Imperial College of Science and Technology,
he was a significant and prolific organic chemist, as well as a major
figure in the academic and social life of the Chemical Society.
Indeed, he did much to raise the profile of the Chemical Society, of
which he was a fellow for nearly seventy years, serving as its president from 1893 to 1895. In 1896, Vernon Harcourt could say
"probably no occupant of the [Presidential] Chair had ever done so
much work for the Chemical Society" as Armstrong had done.
However, in retrospect the twentieth-first-century historian can
hardly admire the successful campaign he waged against the
admission of women chemists to the society.

  In the poorly equipped and ill-lit laboratory of the London
Institution in Finsbury Square, where he was a professor from 1870

until the laboratory was burnt out in 1884, Armstrong and pupils like W. P. Wynne did work of fundamental importance on camphor and naphthalene derivatives. If, in retrospect, none of Armstrong's chemical research proved of any great significance, his many papers were helpful geographical milestones in the understanding of rules of substitution and orientation; while Armstrong's German connections with the firm of Bayer ensured a number of fruitful naphthalene dyestuff patents. One of Armstrong's and Wynne's amino naphthalenetrisulfonic acid derivatives was also later marked out as Fourneau 309, a drug against sleeping sickness.

Despite the confidant bravado of his later life as the elder statesman of science, it is clear that his early life was a struggle. In letters to his father from Leipzig, he often sounds depressed and he expressed regrets that he hadn't decided to be a doctor (having watched Saturday operations at St Bartholomew's Hospitals as a student in London). And although he had the good fortune to fall into a part-time post at Barts (assisting Bunsen's pupil, Augustus Mathiessen, with medical lectures while also helping him in research on alkaloids), as well as occupying the part-time post at the London Institution, he was made despondent by continual failures to secure higher profile positions. In 1874 he was an unsuccessful candidate for the new chair at the Yorkshire College in Leeds; a year later James Dewar pipped him to the Jacksonian Chair at Cambridge; and in 1876 he rejected a chance to teach at Clifton College, Bristol (whose school laboratories launched William Tilden on an impressive career). As late as 1894 we still find him privately insecure when he tinkered with transferring from Central College to the adjacent Royal College of Science— providing he was invited and did not have to apply and face the ignominy of a public defeat. Henry Roscoe's invitation to go to Manchester in 1896 was not followed up for the same reason.

Armstrong was therefore destined to remain a Londoner, albeit one with strong European connections. His great opportunity came in 1878 when the City of London and its Guilds launched a programme of evening class technical education in the basement laboratories and workshops of the remarkable Cowper Street Middle Class School in Finsbury. It was from this that England's first technical college, Finsbury College, emerged as well as the important educational partnership between Armstrong and the electrical engineer, William Ayrton, the mechanical engineer and mathematician, John Perry, and the laboratory designer and architect, E. C. Robins. From their cooperation, as well as Armstrong's earlier teaching experiences, arose the doctrine of heurism and an educational experiment in basing science teaching on finding things out for oneself rather than on teacher instruction and lecture demonstration. Armstrong's first lecture on this approach, given at the International Health Exhibition in the unfinished shell of Central College in 1884, launched him on his career as an educationalist.

We can well wonder at the man's energy in combining a full day of morning and evening teaching with research on organic chemistry, attending and commenting at meetings of the Chemical and Royal Societies, while writing reports on educational subjects for the British Association, the Headmasters' Association and the government, running courses for science teachers, and persuading Dulwich College and Christ's Hospital to practice heurism; and finally, at weekends, playing heuristic games with his own younger children, which they reported meticulously in their notebooks.

All this was done against the background of a career at the City & Guilds Central College which the Guilds had erected at South Kensington in 1884. Here they joined the government's former School of Mines which had also moved to a brand new building in South Kensington, and under Huxley become the Normal School of Science with its own Chemistry Department under Frankland and later T. E. Thorpe. Armstrong became Professor of Chemistry on the third floor of Waterhouse's Central Institution building which had lift shafts but no lifts because Ayrton wanted electrical ones while Unwin, the professor of mechanical engineering, wanted a hydraulic mechanism. Armstrong is said to have merely asked for "a hypothetical lift, please". In fact, such indecision, chronic under-financing of the enterprise, as well as the development of the emergence of petty rules and bureaucracy from the administration, was to make life in South Kensington difficult, and the emergence of a genuine school of chemical engineering well nigh impossible— even if Armstrong had really had the vision to carry this out.

ARMSTRONG'S RESEARCH SCHOOL

Chemical engineering in the later post First World War sense, of unit process teaching based upon physical chemistry, and with strong links with chemical industry, was never realized in Armstrong's day. In any case, we have already noticed Armstrong's aversion to physical chemistry. What Armstrong called "chemical engineering" (and what he liked in the later 1920s and 1930s to regard as preparing the way for what had come to pass) was little more than a combined studies course in mechanical and electrical engineering, mathematics and organic chemistry with a strong dose of heurism in the first year. Instead, what Armstrong developed at Central College was a powerful and influential school of pure chemical research. In retrospect, we may thank Armstrong for this, because it was Armstrong's underlying interest in the mechanisms of chemical change (what he termed "reverse electrolysis"), intramolecular rearrangements, etc. that were to inspire his pupil Arthur Lapworth towards an electronic theory of mechanism. That, in turn, though Armstrong was to disapprove, was to trigger Robinson and Ingold in the 1920s. Armstrong's enthusiasm for the exploration of the relations between composition and crystal form was also bolstered by the advent of X-ray analysis in 1912.

But none of this was necessarily quite what the Guilds had had in mind for its new chemistry school. His employers were, quite rightly, disturbed that while the numbers of students wishing to read three years of mechanical or electrical engineering were buoyant, the numbers taking the three-year chemical engineering route were under a dozen each year. Of course, the reasons for this are and were clear-cut: firstly, the Central College courses were more expensive than, say, University College London; secondly, employers were still not interested in the unknown species of chemical engineer when they could obtain a competent analyst from any of a dozen schools of chemistry across the UK; and thirdly, with the growing interest in candidates taking and obtaining University of London degrees, for which Armstrong was, on principle, opposed to training his students, it did not look like efficient teaching when over half of Armstrong's students failed their BSc. It looked even worse when, after 1900, the University had become a teaching as well as examining institution, and R. B. Haldane and others had began to campaign for both the Royal College of Science and Central College to become part of the federated university.

Hence when in 1907 both colleges (as well as the School of Mines) became constituent parts of the redesignated Imperial

College of Science & Technology, and given Armstrong's age of sixty, the Guild authorities began to look at the possibility of easing Armstrong out. Obvious economies could be made if chemistry and mathematics were transferred and amalgamated with the existing departments in the Royal College of Science, while the Central devoted itself exclusively to engineering. The fact that Armstrong, as well as his fellow nuisances, John Perry and the mathematician, Olaus Henrici, were adamantly opposed to Central College losing its independence and becoming part of the university was only one further nail in their collective coffin. The contracts of Armstrong and the other professors had always been annual ones; hence in 1911, it was easy for the Guilds to say that they would not renew their contracts in 1912. His present and former students were furious and organized what must have been the largest chemical dinner in honour of a single chemist at the Cecil Hotel on 13 May 1911. There were 250 guests.

In point of fact, students were allowed to complete their courses and theses with Armstrong, so that it was not until 1913 that he finally left the Central. Even then, through the kindness of Tom Thorpe and H. B. ("Dry") Baker, he was given bench space at South Kensington which he continued to enjoy until the middle of the war.

Although he coined the Popper-type aphorism, "hypotheses, like professors when they are seen not to work any longer in the laboratory, should disappear", Armstrong did not. Instead he found a new career as a writer and critic, while carrying on good work on enzyme chemistry with his son, Edward.

It is intriguing that Armstrong's criticisms rarely annoyed chemists but did irritate educationalists. (For example, J. J. Findlay, a prominent Edwardian figure in educational circles, was furious in 1914 at what he thought was Armstrong's and Perry's dishonour in running down Britain's education system when the British Association for the Advancement of Science met in Australia.) And perhaps some of Armstrong's thirty-year long critique has rubbed off on historians who continue to blame Britain's economic decline on a failure of education, if not on a failure of South Kensington.

The tone of his retirement writings was often Carlylian and Ruskinian, and therefore biblical—hence my title reference to an epistle. Armstrong's favourite biblical passage was St Paul's Epistle to the Thessalonians—"hold fast to that which is good and prove all things". The infamous 1925 *Journal of Chemical Education* essay, "The Epistle of Henry the Chemist to the Uesanians", attacks Americans for failing to uphold this precept by allowing the physical chemist and editor of the *Journal of Physical Chemistry*, W. D. Bancroft, and others free rein with ions. Armstrong had been to America twice and had returned unimpressed!

THE CHEMIST IN THE COLOURED WAISTCOAT

In the 1930s, when he was already in his eighties, Armstrong started a fashion for coloured waistcoats. The practice arose when he helped to organize the Faraday electromagnetic centenary celebrations in the Albert Hall in 1931. Appalled to find that the Institution of Electrical Engineers planned to ignore Faraday, the chemist who had discovered benzene in 1825, he hung a "Sign of the Hexagon" in the organ gallery with what appears to have been a marvellous display of dyestuffs. Sensibly, the Royal Institution's managers arranged for him to give a Friday Evening Discourse on the same occasion—and it was at his evening discourse that he stunned the audience by wearing a sky blue and pink waistcoat with his formal evening dress, as well as arranging for a group of ballet dancers to wear (and shed!) a stunning series of coloured gowns and fabrics. In the following year, 1932, full of social bonhomie as ever, he organized a dinner-dance for London chemists at which all the male guests were directed to wear coloured waistcoats.

Armstrong's point was that chemists should be proud of the industrial and commercial applications of their brain power—an echo, here, of a famous lecture given by A. W. Hofmann's at the Royal Institution in 1865 in which he rhapsodised on the transmutation of dirty coal into beautiful dyestuffs.

Finally, then, what of Armstrong's epistle, preached for fifty years from the mid-1880s? Let me quote from a 1929 Carlyle-inspired talk, 'The Future Allchemist' (*sic*):

"In 1874 I published an elementary textbook on *The Chemistry of the Carbon Compounds* which was a well-proportioned, fairly articulate skeleton of the subject, carrying sufficient clothing to warrant its appearance in decent society. A vast amount of extra clothing has been piled upon that skeleton in the interval—numberless petticoats and frills from here, there and everywhere; nevertheless, there are very few bones of moment to add to that fifty-year old framework. I am satisfied that a similar fair presentation of the subject, as it stands today, might be given, without greatly increasing the length of the book."

Then comes the punch line:

"Women have set an example by showing how little clothing it is necessary to wear—how much more freely and advantageously the skeletal mechanism can be used when lightly clad."

No Ingold or quantum mechanics for Armstrong! Armstrong's epistle was that chemistry had become cumbersome and that the blame lay with physical chemists intervening in elementary teaching. Tyros no longer thought things out for themselves; the result was badly trained chemists, with bad results for industry. Armstrong's very distinctive voice and approach to chemistry teaching and research is encapsulated in one of his aphorisms. "On any mountain, the route of the historic first ascent is almost invariably still the easiest route to the summit". In other words, the easiest approach to teaching and to understanding a scientific problem was *via* the historic first ascent. Small wonder, then, that the historian of chemistry finds Armstrong a sympathetic character.

Armstrong's personality was elegantly summarized by his son-in-law, Stephen Miall, the editor of *Chemistry & Industry*. "He was a vigorous man physically and mentally, a keen critic, a great worker, a good friend, and a good companion, a veritable Dr Johnson of modern times, with many of the good qualities, and some of the faults characteristic of the great lexicographer".

CHAPTER 21

He Knew He Was Right—Fritz Haber

In May 2004 British television viewers were regaled by an excellent dramatization of *He Knew He Was Right*, the *Othello*-like tragedy published by Anthony Trollope in 1869 about a jealous husband driven insane by his innocent wife's supposed infidelity. While a guide to how Victorian husbands may have behaved in the 1860s when men were always in the right, the reading today is that the obsessed husband was certainly wrong. We also have tragic heroes in science whose success seems fatally flawed. None more so than in the German physical chemist Fritz Haber who, as chemist, Christian–Jew, husband and patriot always knew he was right.

Haber is a paradigm for all human life, illustrating how every advance in human understanding can be used for good or evil, and how consequences can never be entirely foreseen. On the one hand, his laboratory synthesis of ammonia from its elements, when scaled up industrially by Carl Bosch, made the prevention of mass starvation possible. On the other hand, at the moment of its industrial implementation by BASF, it enabled the Second Reich to produce nitric acid for the manufacture of conventional explosives during the First World War when deprived of South American nitre by British naval blockades. It was Haber himself who promoted this change of direction.

The Case of the Poisonous Socks: Tales from Chemistry
By William H. Brock
© William H. Brock, 2011
Published by the Royal Society of Chemistry, www.rsc.org

As the Nobel Prize-winning chemist Roald Hoffmann has depicted in a poem, Haber saw himself as a human catalyst who would enable the war to be brought to a rapid conclusion. Convinced that Germany was right to fight a European war, and appalled by the stalemate of trench warfare, Haber persuaded military chiefs that British and French soldiers could be forced to run from their dugouts by blowing chlorine gas at them. This was in contravention of the Hague Convention of 1899 that had prohibited the use of asphyxiating weapons. He was actually present to direct the initial operations during the first gas attack at Ypres in April 1915. The reaction was inevitable and all escalated horribly, with Haber put in charge of investigations of new weapons such as phosgene and mustard gas as well as preventative procedures.

He could never understand the anger of French and British scientists at his award of the Nobel Prize for Chemistry in 1918 for the ammonia synthesis that had prolonged the war. It was also a source of humiliation for Haber that, despite his patriotic transformation of his research institute into a military research centre, the German High Command gave him the rank only of Captain, whereas his equivalent number in Britain, the physical chemist Harold Hartley, was promoted to Brigadier-General.

Haber remained convinced that chemical warfare was no worse than incendiary or explosive weapons. Consequently, under the guise of research on pesticides he directed secret work on chemical weapons throughout the 1920s—weapons that were bought (and used) by both the Spanish and Russian governments. One of the most effective agricultural pesticides was hydrogen cyanide and Haber was directly responsible for the Zyklon process whereby it could be prepared safely on a large scale. Might he not have foreseen its use to kill thousands of human "pests", as the German hygienists perceived Jews and other outsiders? (Figure 21.1).

Haber was born in Breslau, now Wroclaw, in East Prussia of Polish Jews who had worked in the wool trade for centuries. His father, a dyer, encouraged his son to study chemistry, probably with a view to employing the new synthetic dyestuffs that were replacing traditional natural dyes such as madder and indigo. He was therefore trained as a typical *Organiker* before undertaking a variety of short-term business experiences in chemical works in Austria and Hungary. However, when he eventually joined his father's business, it soon became evident that he was not cut out for

Figure 21.1   Fritz Haber (1868–1934), German physical chemist sketched by
Wilhelm Luntz in 1911 when he became director of the Kaiser
Wilhelm Institute in Berlin. (Brock)

commerce and within a few months he left home for further chemical training with Ludwig Knorr at the University of Jena. Here,
in order to smooth his path to an academic position, he converted
to Christianity. He then spent seventeen fruitful years at the
Karlsruhe Hochschule where he also converted himself into a
physical chemist through his friendship with Hans Luggin, a pupil
of Arrhenius's. He once told Staudinger that traditional organic
chemistry was finished unless it began to use physics—a prescient
suggestion that most European organic chemists ignored until after
Haber's death.

   The years at Karlsruhe were immensely productive and made
glorious by Haber's close concern with chemistry's applications.
J. D. Bernal rightly declared Haber to have been the world's
greatest authority on the relations between scientific research
and industry. It was at Karlsruhe that he made fundamental contributions to electrochemistry and the thermodynamics of gas
equilibria. From the latter he demonstrated that the synthesis of
ammonia from its elements was economically viable if a catalyst
and high pressures were applied.

Haber was deeply ashamed that Germany had lost the war and he did everything in his power to re-establish her hegemony in the world's science. Faced by Germany's appalling post-war debts he initiated a secret project to recover gold from the sea. Although the project was a failure, it led to important improvements in analytical techniques. More successfully, he strove to bring German science back into the international fold by holding innovative colloquia at Berlin's Kaiser Wilhelm Institute for Physical Chemistry that he had directed since 1911. All his efforts in this direction were destroyed in 1933 when the Nazis began to dismiss Jewish civil servants from their positions. Already a victim of angina from the cigars he incessantly smoked, and the lonely survivor of two marriages broken by his obsessive work habits, he died in exile in Switzerland at the end of 1933.

Although Haber is a principal character in Tom Harrison's extraordinary music drama *Square Rounds* performed at London's Olivier's Theatre in October 1992, it is odd that playwrights and screenwriters have not seized upon the dilemmas and contradictions of his chemical career to create an epic drama. The Austrian film actor and director Erich von Stroheim, memorable as an arrogant Prussian officer in von Stroheim's *The Blue Angel* (1930) and Renoir's *La Grande Illusion* (1937), bore an astonishing resemblance to Haber and would have been perfectly cast in a 1930s film following Haber's death. One can easily imagine von Stroheim playing Haber as he delivered his postponed Nobel laureate speech in 1920: "In no future war will the military be able to ignore poison gas. It is a higher form of killing".

Haber was a chemist who did not always get it right. He was of towering stature in the social history of early twentieth-century chemistry and as a research director obsessed by work.

CHAPTER 22

J. R. Partington (1886–1965): Physical Chemistry in Deed and Word

Today James Riddick Partington (1886–1965) is remembered as an historian of chemistry rather than as the significant British research chemist and textbook writer he was perceived to be in the 1920s and 1930s. Because his textbooks were specifically geared to the British secondary school and university systems, he is probably not well-known in the United States as a textbook writer. Nor in America or Europe is he remembered as a practising physical chemist who made contributions to thermodynamics, the determination of specific heats, and to electrochemical theory. So, for example, he is unmentioned in Keith Laidler's *World of Physical Chemistry* (1993). Nevertheless, as an outstanding example and model of the chemist–historian, it is of interest to examine his career as a chemist.

EARLY CAREER AND ESTABLISHMENT AS A LONDON CHEMIST

Partington was born on 30 June 1886 in the tiny coal-mining village of Middle Hulton to the south of Bolton. His father was a book-keeper in Bolton and his mother, from whom he took the middle name of Riddick, was the daughter of a Scottish tailor. While still quite young his parents moved to the seaside town of Southport, to

the north of Liverpool, allowing Partington the benefit of education at the Victoria Science & Art School which had opened in
1887. Here his prowess as a mathematician and practical chemist
must have been forged. He left school in 1901 when he was fifteen
because his parents moved back to Bolton. There he began to assist
the town's Public Analyst, a post that must have involved the
acquirement of the skills in volumetric and gravimetric analysis
that were a hallmark of his later work. After a couple of years, and
still in local government employment, he became a lab assistant in
the town's Pupil Teachers Training College before finally becoming
a clerk in Bolton's education offices.

During these five years between 1901 and 1906 he embarked
upon an intensive course of part-time private study, developing his
knowledge of foreign languages, and mathematics. In 1906, at the
age of 20, he qualified for entry to the University of Manchester to
read chemistry and physics. He would have used the laboratories
that Henry Roscoe had erected in Oxford Road in 1872. Among his
teachers were Harold Baily Dixon (1852–1930), "whose lectures",
Partington recalled, "were illustrated by striking experiments, were
brilliant, stimulating, and in close contact with original sources and
research. They were sometimes enlivened by touches of his characteristic humour". He was, however, "somewhat hampered by
insufficient knowledge of mathematics". In other words, although
it must have been Dixon who taught Partington thermodynamics,
the pupil felt he knew more than his teacher. Other instructors
included W. H. Perkin Jr, but Partington was never interested in
organic chemistry. An interest in the history of chemistry was
engendered both by Dixon and Andrew Norman Meldrum (1876–
1934), whose Carnegie Research Fellowship overlapped with
Partington's undergraduate and postgraduate studies. Meldrum
had already published an outstanding study of the atomic theory in
1906 and was planning to write a history of chemistry in his
spare time. Although he emigrated to India in 1913, Meldrum and
Partington remained in close contact.

On graduating in 1909 with first class honours and being granted
a teaching diploma, Partington was awarded a fellowship funded
by the Manchester engineering firm of Beyer to begin postgraduate
research with the physical organic chemist Arthur Lapworth,
whose first research student he was. Astonishingly, within a year
he had published two papers in the *Transactions of the Chemical*

Society and a further four in 1911 before gaining his MSc. The first paper, written with Lapworth, confirmed that the presence of water in the hydrolysis of an ester diminished the catalytic influence of hydrogen chloride. In the second paper he investigated ionic equilibria in electrolytes from a thermodynamic viewpoint. Effectively, this was a study of the literature on Ostwald's dilution law and the reasons strong electrolytes diverged from the law of mass action. Both these early papers show Partington adept at thermodynamic reasoning and his commitment to research in the area of electrolysis, as the other four papers confirm. This was, by any measure, an astonishing output from a postgraduate student of twenty-four.

Then, even more astonishingly, in 1911, and while still a graduate student, he published his first textbook on *Higher Mathematics for Chemical Students*. Nernst and Schönflies had published the first "math for chemists" text in 1898 and this had appeared in English in 1900. Partington gave no reason for publishing his textbook and this is odd, given that John William Mellor, a previous student of Dixon's (and with his ardent support) had published *Higher Mathematics for Students of Chemistry and Physics* nine years earlier in 1902. Longmans had kept this in continuous print, so why the need for Partington's book? His dense introduction on scientific method, which shows him already very familiar with the history of chemistry, provides no clue. All one can say is that Partington's text was shorter (272 pages) than Mellor's (600 pages) and that it was less detailed. Both texts remained rivals and in print until the Second World War, following which Partington re-used much of the material as the introductory chapter of the first volume of his multi-volume treatise on physical chemistry. Little wonder, then, that a reference from Dixon, refers to Partington as "one of the most brilliant students we have had during the last thirty years".

Armed with his MSc in 1911, Partington went to Berlin to study with Walther Nernst, though for reasons unknown, he did not complete a PhD. When he arrived, he spoke German imperfectly, but was soon asked to give a seminar. He carefully wrote this out to read so as not to stumble, but Nernst kept interrupting, forcing Partington to speak without a script. This was Nernst's way of giving him confidence! Following the deduction of his heat theorem in 1906, Nernst had urged chemists to undertake a programme of experimentation on the heats of reaction, specific heats and temperature coefficients to test whether the theorem was an approximation to truth or a true third law of thermodynamics. This gave Partington his programme of research in physical chemistry for the next thirty years; the testing of theory against very precise physical measurements.

Partington stayed in Berlin until 1913 working on the variations of specific heats of gases with temperature using an adaptation of the adiabatic expansion apparatus first developed at the University of Berlin by Otto Lummer and Ernst Pringsheim. He had to persuade Nernst that an improvement of the Berliners' complicated apparatus was needed, since Nernst "had a profound distrust of large, complicated, and expensive apparatus". Nernst refused to speak to Partington for a couple of days before relenting and providing him with his own resistance box and string galvanometer for the experiments. Partington used the change in resistance of a Wollaston platinum wire as a thermometer. The wire was placed in a copper balloon with a capacity of 130 litres and the gas expanded through a stopcock.

Although there was to be no Berlin PhD, Partington did publish five papers in German on his research in the Leipzig journal, *Physikalische Zeitschrift*. These were on the specific heats of air, carbon dioxide and chlorine, and on heats of vaporization and evaporation. While in Berlin he must also have drafted his next book on *Thermodynamics* since it appeared immediately on his return to England in 1913. The text was indebted to the insights of Nernst's *Theoretische Chemie* (1893) which had been translated into English in 1907. A detailed account of classical thermodynamics, the last two chapters dealt with Nernst's heat theorem and with energy quanta. *Nature* thought it tough reading for chemists unequipped with mathematics. Partington later described his thermodynamics text as "a pioneer work, [as] nothing of its scope and character was then available in English". This was true since Lewis and Randall's influential textbook did not appear until 1923 and the only major competitor was Nernst's.

Not surprisingly, Partington had been welcomed back to the University of Manchester in 1913 as a lecturer. One of his first students was Marian Jones, the daughter of a brickworks manager from Chester whom he supervised for an MSc degree on supersaturated solutions. Partington fell in love with his student and married her after the war on 6 September 1919. She became a chemistry schoolteacher before having two daughters and a son, Roger, who also became a physical chemist.

Immediately war broke out in 1914, Partington joined the army, only to be seconded to the Ministry of Munitions to work on water purification with the young physical chemist Eric K. Rideal. Later the two chemists turned to the question of the oxidation of nitrogen to form nitric acid and investigated the Haber–Bosch process that the Germans were pursuing. This led to a book on the alkali industry in Rideal's series on the chemical industry in 1918 and later collaboration with Leslie Henry Parker on a history and analysis of the contemporary post-war nitrogen industry. For his war work, Captain Partington was awarded the MBE (Military Division). Outside his war work for the government, Partington managed to continue with thermodynamics, joining the Faraday Society in 1915. In 1919 he presented a major review of the literature on the dilution law to the Faraday Society to whose Council he was elected that same year.

In 1919 he was appointed sole Professor of Chemistry at the East London College (it was renamed Queen Mary College in 1934). This Victorian enterprise had begun life as the People's Palace in 1887 as a place of entertainment and education for the poor living in the insalubrious conditions of London's East End. Its educational functions rapidly became more important than its leisure ones and it was recognized by the University of London for degree purposes in 1915. Partington's immediate predecessor as Professor of Chemistry was John Hewitt (1868–1954), an organic chemist whose pupils had included Samuel Glasstone. Hewitt had designed a three-storey laboratory building in 1914 and Partington subsequently added a fourth storey in 1934 (Figure 22.1).

The conditions for teaching and research were hardly ideal. Accommodation and discipline were serious problems because of the influx of men from war service, and laboratory stocks of chemicals and equipment were dire. However, with the support of the College's administrators, and with small grants from the Chemical Society and the Department of Scientific and Industrial Research, Partington succeeded in establishing a modest research school with colleagues such as W. H. Patterson and D. C. Jones, and on the organic side, F. G. Pope, E. E. Turner and H. D. K. Drew. College calendars show that that Partington's rate of publication not only

Figure 22.1 James Riddick Partington (1886–1965) in his mid-forties while professor of chemistry at Queen Mary College. There he developed a reputation as a physical chemist, textbook writer and, increasingly, an historian of science. (Science & Technology Picture Library)

outshone that of his chemistry colleagues, but those of colleagues throughout the college. Even so, when Michael Dewar inherited the Department in 1951 he complained at its shabbiness and unsuitability for research, but the chemistry building was not demolished and rebuilt until 1967 after Dewar had left. Partington chose to lecture exclusively on inorganic and physical chemistry. A compulsory one-term course on the history of chemistry, which he introduced in 1919, was soon abandoned, though he revived it as an elective from 1945 onwards.

With the outbreak of the Second World War in 1939 Partington's department was evacuated to Cambridge and Partington spent the war years in that city enjoying the facilities of the university's copyright library. Although arrangements had been made for the families of staff to be accommodated at Cambridge, Mrs Partington stayed behind at the family home in Wembley. Tragically, she committed suicide in March 1940, leaving Partington a widower for the remainder of his life.

On returning to the badly damaged East End of London in 1945 he more or less abandoned laboratory research and devoted himself instead to historical work and to the completion of his *Advanced Physical Chemistry*. He retired in 1951 to a house in Mill Road, Cambridge, where he was looked after by an aged housekeeper. The house was filled with books from cellar to roof. According to Joseph Needham, he became something of a recluse, rarely stirring from his writing desk. At the end of 1964, following his housekeeper's retirement and unable to look after himself, he joined relatives in the salt mining town of Northwich in Cheshire where he died on 9 October 1965.

PARTINGTON'S RESEARCH IN PHYSICAL CHEMISTRY

Throughout the 1920s and 1930s, Partington made many other contributions to Faraday Society discussions. Although never elected president (probably because his modesty and intense reserve deterred him from seeking such office), he served on Council almost continuously from 1919 to 1938, and particularly on its Publications Committee on which he also served as the representative for the American *Journal of Physical Chemistry*.

Partington's 1919 Faraday Society paper was a critical examination of theories of strong electrolytes. In particular, it examined Jnanendra Chandra Ghosh's theory of strong electrolytes published the previous year and showed that it was not in agreement with experiment. Ghosh assumed complete dissociation of strong electrolytes and that the majority of the dissociated ions arranged themselves into a crystal-like space lattice. Partington found the theory "startling" but deduced that it was incompatible with observed data. Ghosh, who was due in England to take up a research post at University College London, was not present but sent in a reply. Unfortunately, Partington made an arithmetical blunder that enabled Ghosh to rebut the valid criticism that Partington had made. Partington's response showed again that Ghosh's theory was based upon "guesswork". According to an appraisal of Ghosh by R. Parthasarathy in *The Hindu* on 12 December 2002, the criticism caused Ghosh to withdraw from being elected Fellow of the Royal Society! This is obviously based upon a misconception but may, perhaps, have been an anti-imperialist story told by Ghosh in later years.

Partington's other principal research was on the temperature dependence of specific heats. As we have seen, this interest was initiated by Nernst while Partington studied in Berlin. Once settled at Queen Mary College, Partington took up this research again. Whereas Nernst had been interested in the determination of specific heats at low temperatures because of quantum effects, Partington was interested in their behaviour at high temperatures. There were obvious industrial applications in the automobile and refrigeration industries, as well as the need for specific heat data in designing industrial plants involving gases. Instead of measuring specific heats by adiabatic expansion, as he had in Berlin, he determined c_p/c_v from the velocity of sound using a modified Kundt tube, as Dixon had recently done at Manchester. He initially determined values for air and some simple gases using a modified electrically heated Kundt tube to determine the velocity of sound at different temperatures. Later, with W. G. Shilling, the son of the owner of an engineering firm, he developed a modified and improved form of the apparatus to enable measurements up to 1,000 °C. The joint work was summarized in 1924 as a "coherent and critical account of the state of our knowledge".

Work on specific heats led Partington into an interesting controversy with the young Mrs Ingold. In 1921 Hilda Usherwood, the future wife of Christopher Ingold, working with Martha Whiteley at Imperial College, began an investigation of tautomerism using the variation of specific heats with temperature as a guide to changes of equilibria. Her two papers on "the detection of tautomeric equilibria in hydrocyanic acid" and "specific heats of gases with special reference to hydrogen" (the latter with Ingold) appeared in 1922. In 1925, a year after he and Shilling had published their book, *The Specific Heats of Gases*, Partington challenged Mrs Ingold's results. He claimed her values for hydrogen had been only approximate, that her HCN was impure, and that her values for the hydrogen cyanide–hydrogen isocyanide equilibrium were due not to thermal effects accompanying isomeric change, but polymerization, which Hilda Ingold had ignored. She replied, standing her ground; and Partington stood his. But Mrs Ingold won the day by showing that Partington's evidence for association was valid only for a very small part of the temperature range studied. In his biography of Christopher Ingold, Kenneth Leffek suggests that Partington's criticisms were weak and that "in 1925 Partington felt that it was fashionable to attack someone with the name Ingold, in view of all the activity in the Chemical Society and in the pages of *Chemistry & Industry* concerning the theory of chemical reactions". This is unfair. Partington's 1925 paper, based upon an MSc thesis by his pupil M. F. Carroll, merely noted that measuring the specific heats of HCN by a different procedure to Mrs Ingold gave different results and suggested why this might be so. It is clear, in any case, that the Ingolds did not hold the controversy against Partington since Christopher Ingold signed Partington's Royal Society application in 1926.

PARTINGTON AND THE ROYAL SOCIETY

Partington had read three papers on specific heats to the Royal Society in the years 1921–1925 and these had been communicated by Dixon and the physicist, J. A. Harker. He was first put up as a candidate for its Fellowship in 1927 during the Presidency of Ernest Rutherford. By 1924 when the book on specific heats appeared, Partington had published some eleven papers on specific heats and could be considered the British expert on the subject. Given Partington's publication record and his prominence in the Faraday Society, why was his candidature a failure?

In the 1920s, election to the Fellowship was by recommendation in writing by six or more Fellows, of whom three had to be recommending from personal knowledge. A printed list of candidates was circulated to all the Fellows each January. The Society's Council then selected twenty of the names by ballot and recirculated its proposals which were then voted on by those Fellows present at the next ordinary meeting. Proposals were allowed to stand for four further years after first time failure, following which the candidate could be proposed again by new sponsors. Having failed to have Partington elected the first time in the years 1927 to 1931, he was proposed a second time from 1935 to 1939. The first two signatories were conventionally understood to be the proposer and seconder, and in Partington's case they were the physical chemists Herbert Brereton Baker and Frederick George Donnan in 1927, and Eric K. Rideal and Donnan in 1935. All three sponsors had connections with Partington through his wartime activities and were prominent in the affairs of the Faraday Society.

However, Baker and Donnan made a poor job of the nomination, merely stating that Partington was "distinguished for his research work in inorganic and physical chemistry" and citing a few papers (but missing out his many contributions to the Faraday Society) and saying that there were 52 other papers as well as books on thermodynamics, inorganic chemistry, mathematics for chemists, and five other books. Despite this lack of specificity, the nomination attracted many distinguished chemists, including Nevil Sidgwick, C. K. Ingold and William Jackson Pope.

That proposal having failed, Partington was sponsored again in 1936 during the Presidency of William H. Bragg. This time the sponsors, led by Rideal and Donnan, were more elaborate in extolling Partington's virtues as a scientist.

"The candidate has published numerous scientific papers and several valuable text books since 1910. Of the latter, one on higher mathematics for chemical students, the other on inorganic chemistry are in their fourth edition, and one on thermodynamics is in its second edition. His work on the specific heat of gases by classical methods is well known, and several of his determinations are accepted internationally. He has also published two series of papers, one on dielectric polarization and the other on concentration cells which are records of careful and accurate work in physical chemistry. He has investigated analytically a number of unusual inorganic reactions and elucidated their mechanisms. There have been published in the *Journal of the Chemical Society* and the *Transactions of the Faraday Society*. His interests in the history of chemistry are exemplified by a series of papers and a research monograph of unusual character."

As both proposals show, my initial assumption that being a writer of textbooks and history of science counted against Partington does not seem to have been the case. On the other hand, Partington's research was hardly innovative; rather it relied upon perfecting others' work, or what T. S. Kuhn aptly described as "normal science". Partington was not blazing any new trails in his research such as those being undertaken in the 1920s in quantum chemistry, kinetics and spectroscopy.

A comparison with Joseph Mellor, another encyclopedic chemist, is especially apt since he was one of the two chemists elected in 1927 in preference to Partington. Mellor was also largely self-taught before gaining his first degree at the University of Otago in New Zealand by part-time study. Like Partington, he had then joined the University of Manchester, where he wrote his previously mentioned text on mathematics for chemists and his *Chemical Statics and Dynamics* (1904). Unlike Partington, however, he did not become a university teacher; instead he used his deep knowledge of physical chemistry to transform the ceramics industry of

Staffordshire. Although, like Partington, he continued to publish excellent textbooks on inorganic chemistry, including the multi-volume *Comprehensive Treatise on Theoretical and Inorganic Chemistry* (1922–1937), it was the originality of his research in ceramics chemistry, where he opened up an economically important industry to scientific scrutiny that brought him the Fellowship of the Royal Society (FRS) in 1927. Similar points can be made about originality for all the many other chemists who were successfully elected FRS between 1917 and 1939.

CONCLUSIONS

Throughout the 1920s and 1930s Partington regularly published five or six papers a year either independently or with students on a variety of topics in inorganic and physical chemistry. All his work was characterized by meticulous experimentation and the gathering of quantitative information whenever possible. The whole of Partington's research was devoted to the appraisal of deductions made from thermodynamic equations and comparison between theory and experiment with the aim of perfecting theory and the creation of sound and accurate physical constants and measures. For example, he did lots of work on solubility effects, and devised and developed a new form of electric vacuum furnace in 1925 to investigate high temperature reactions.

Partington had well over 70 collaborators between 1914 and 1951 (when he retired). Among his pupils were Frederick E. King, later a professor at the University of Nottingham before he entered the chemical industry; Arthur Israel Vogel, the analytical chemist and textbook writer; and Raymond J. W. Le Fèvre, who was not impressed. It was said of Harold Dixon that he was singularly reticent and was "difficult to penetrate within his outer ring of electrons". The same was true of Partington, though one obituarist, thought him reserved rather than reticent and that he was "extremely modest". He was a small man with a military bearing and teutonic and seemingly testy in manner. Conservative in dress, he still wore a wing collar until quite late in life. He spoke very quietly so that students and fellow academics often found his lectures inaudible, and therefore boring.

His working methods were those of the Victorian and Edwardian scholar. He wrote neatly (or typed) on the backs of proofs which he then cut up and rearranged as necessary by gumming them together. Patient printers and publishers allowed him to tinker with several proofs until he was satisfied with their accuracy. His encyclopaedic four-volume *Physical Chemistry* (1949) was compiled at Cambridge during the war and kept in a suitcase which he carried into underground air raid shelters to work on during German air raids.

Partington was a highly competent practical and theoretical chemist, and gifted (as Hartley remarked in the *Dictionary of Scientific Biography*) with an encyclopaedic mind. But although the problems he tackled were often intricate, they could be rather dull normal science. He seems to have lacked the ability, or the desire, to tackle frontier problems. Undoubtedly he gave excellent training to several generations of chemists (including several from India) who went into teaching or industry, while his texts offered great value to generations of school and university chemistry students. Nevertheless, just as his four-volume *History of Chemistry* (1961–1970) is an indispensable aid to historians of chemistry, his chemistry papers, his *Higher Mathematics for Chemical Students,* his *Thermodynamics*, his *Specific Heats of Gases*, and his huge *Advanced Physical Chemistry* remain monuments to the development of physical chemistry since the 1900s. What Partington wrote of Nernst in 1953 is equally a memorial to his own work as a physical chemist.

"A physical chemist is at some disadvantage, compared with the organic chemist, since new compounds remain, but new [physical] measurements soon give way to newer, and sometimes better, ones. The pioneering investigations are soon forgotten, and results which in their time were highly important and significant are amplified and revised by later workers, who not infrequently reap the benefit of newer techniques which make their task easier than that of the earlier pioneer experimenters, whose contributions to science tend to be overlooked."

CHAPTER 23

Henry Crookes, Founder of Crookes Laboratories

The wonderful collection of paintings at the Barnes Foundation at Merion in west Philadelphia was created by the American chemist and pharmacist Albert C. Barnes (1872–1951) who had made his fortune from sales of the bactericide Argyrol, a mixture of silver oxide and protein. Mild solutions of silver nitrate had been used in the prevention of ophthalmia in new born babies in America since 1884. Barnes had developed his silver-based colloidal bactericide, which proved particularly valuable as eyedrops, with a German–American chemist, Hermann Hille, in 1902. By 1907 international sales reached more than four million bottles, with net profits of over $175,000. But by then the stormy partnership between Barnes and Hille had broken down and Barnes bought Hille out, making himself a millionaire. Later, in the 1920s, in exchange for patents, Barnes formed a partnership with the British firm of Crookes Colloids, to form Crookes–Barnes Laboratories whose factory was at Wayne, New Jersey. In 1929, preferring art and education to pharmaceutical manufacturing, Barnes sold his business to the American antiseptics manufacturer, Zonite Products Corporation, in exchange for six million dollars' worth of stock. At about the same time, back in England, Crookes Colloids was transformed into Crookes Laboratories with a large modern factory at Park Royal in west London.

The Case of the Poisonous Socks: Tales from Chemistry
By William H. Brock
© William H. Brock, 2011
Published by the Royal Society of Chemistry, www.rsc.org

The Crookes referred to was Henry Crookes (1859–1915), the eldest son of the famous chemical physicist, Sir William Crookes (1832–1919). William and Ellen Crookes had six sons, all of whom were expected to follow a career in science and business. Like Henry, Joseph Crookes (1861–1902) and John Crookes (1863–1931) were educated at University College School in Gower Street. Following some preparatory education in Uxbridge, the three youngest sons, Bernard (1865–1930), Walter (1867–1935) and Lewis (1874–54) went to the City of London Boys School before completing their education at Felsted, the famous public school in Essex. In the event, John, Walter and Lewis showed no interest in science whatsoever. Walter became an accountant and Lewis a solicitor. John, the black sheep of the family, became a drifter in Canada and was disinherited in his father's will. Joseph and Bernard completed their education at the Royal School of Mines. Joseph took up electrical engineering with the Post Office but died young before making his mark. Bernard, who practised as a civil engineer in Newcastle, while occasionally helping his father with the analysis of data, remained uninterested in pursuing research himself. Much, therefore, was expected of Henry, who attended University College School between 1871 and 1876. This was the period when his father was deeply involved in the investigation of psychic phenomena and Henry was both a witness and participant in séances with the mediums Daniel Dunglas Home and Florence Cook. It is a matter of regret that he never published an independent account of what he saw (Figure 23.1).

To boost Henry's health, which was never robust, his father arranged with William Allen Miller, the Professor of Chemistry at King's College London for him to spend a year in Australia learning assaying techniques with Miller's brother, Frederick B. Miller, at the Melbourne Mint from 1877 to 1878. Meanwhile, back in 1873, when working for the sewage treatment Native Guano Company as its chief chemical consultant, William Crookes had developed a process for dealing with the particularly noxious problem of recycling refuse from the fishmongering and butchering trades. By treating separate portions of the odiferous wastes with acid and alkali and then mixing them together, Crookes was able to separate out the more valuable nitrogenous materials. The latter were then treated with carbolic acid and subjected to mechanical drying and pulverization. The result, he claimed, was the

Figure 23.1   Henry Crookes (1859–1915), the eldest son of Sir William Crookes who spent much of his life in his father's shadow, before developing the medical applications of silver colloids. (RSC Library)

production of a niche-market fertilizer for farmers and gardeners interested in growing plants (potash dominant), root crops (phosphorus dominant) and cereals (nitrogen dominant). Convinced he had a money-making scheme he persuaded one of the directors of the Native Guano Co., the stockbroker A. C. Ionides, to finance the creation of an independent "Crookes and Company" to work and market the process. Works were opened at Rainham in Essex on the shores of the Thames, but repeated problems with unreliable machinery seriously delayed production. When the eighteen-year old Henry Crookes returned from Melbourne in 1878, he was placed in charge. However, by the end of that year, Ionides felt that he had borne enough of the start-up costs and the unprofitable works were closed. It is interesting to reflect that at this date William Crookes had no inhibitions that by associating his name with a commercial venture he might damage his scientific reputation. It was to be a different story thirty years later when it was Sir William's turn to be associated in one of his son's projects.

Following the closure of the fertilizer business, Henry and Joseph were packed off to Paris to study chemistry in Adolphe

Wurtz's private chemical school, where the brothers boarded with an analytical chemist. (William Crookes had published a translation of Wurtz's *An Introduction to Chemical Philosophy* in 1867.) Following a year in Paris, Henry studied mining engineering at the Royal School of Mines from 1879 to 1881, emerging with a certificate in metallurgy. By this date, William Crookes's work on the radiometer had led him (and others like Joseph Swan and Thomas Edison) to begin investigations of the conditions inside evacuated glass vessels and from there to the commercial possibility of making an incandescent light. At an important international exhibition of different kinds of lamp bulbs in Paris in 1882, William Crookes was a juror and he and Henry determined the efficiencies of the exhibits by measuring their resistances and illuminating powers at two different candlepowers. Edison's lamp was declared the winner.

As a juror, William Crookes had been unable to demonstrate his own lamp; nevertheless, on returning to London, father and son set up an incandescent lamp factory at Battersea with Henry as manager. This was the excuse for William Crookes to use the punning Latin tag "ubi crux, ibi lux' (where there is a Crookes there is light) as his business motto. At Battersea, father and son, helped also by William's personal assistant, James H. Gardiner, pursued research on the design and improvement of filaments, including one made from metals. Henry, who together with his brother Joseph had been elected Associates of the Institution of Electrical Engineers in 1883, also experimented with electrotyping using copper and a 150-volt dynamo, but nothing came of this. It was through his father's friendship with the railway electrician C. E. Spagnoletti, and their common membership of the Institution that Henry met the latter's daughter Madalina. Henry married Madalina (known in the family as Nina) in 1883 and settled in a house in Fulham. There were no children. She was a talented artist: together with Henry's sister, Alice Crookes, she prepared the many diagrams and illustrations for William Crookes' brilliant presidential address to the Institution of Electrical Engineers in 1891.

Sales of the Crookes lamps were poor and, badly under-capitalized, the Crookes electric lamp company was unable to manufacture lamps of a large enough scale to compete with the combined forces of Edison and Swan and other smaller, better-capitalized, companies. Reading between the lines it is also possible that with so many irons in the fire, William Crookes had to leave too much to Henry who never seems to have been totally committed or wholly industrious at anything he undertook. If this is too harsh a judgement, perhaps Henry resented being dragooned by his father into a business that he was not really interested in. In 1889, the Battersea lamp factory was abandoned and the assets sold to the Geilcher Lamp Co. in Bishopsgate, for whom Henry managed its incandescent lamps department for a time. Within a few months, however, the Geilcher Co. collapsed as a result of the judgement reached in the *Edison v. Swan* patent case and Henry found himself without an occupation. He now decided to try his fortune in the Transvaal gold fields using his father's improved method of sodium amalgamation, leaving Madalina to live with her in-laws in Notting Hill. Over the next few years Henry held managerial positions at mines in Johannesburg and neighbouring Klerksdorp, as well as spending time prospecting for gold around Mafeking, then part of British Bechuanaland. Unfortunately, Africa did not suit his health and bouts of malaria drove him back to England in the 1890s, when he and Madalina set up home near Harrow. There, like his father, Henry worked as a chemical and mining consultant.

When his father bought some gold and silver mines in central Wales in the 1890s, Henry was placed in charge of engineering operations and assaying. Back in the 1860s and 1870s, William Crookes had been much involved as a consultant to companies that mined gold in Wales by making improvements to his sodium amalgamation process. The development of the cyanide process for gold extraction by J. S. MacArthur and the Forrest brothers (Robert and William) from 1887 onwards gave William Crookes further hope for financial success in Wales. Between 1887 and 1890 Crookes and his sons Bernard and Henry acquired prospecting licenses from the Crown for several potential sites around Dolgellau in Merionethshire and formed the Proprietary Gold Recovery Co. with Henry in charge of operations. Although this family enterprise enjoyed some success in that gold was mined and sold, in the long run expenditure exceeded profits and operations were abandoned in 1895. It must have seemed as if every job Henry tackled ended in disaster.

Henry returned to London and settled at 109 Ladbroke Grove, close to his parents' splendid house in Kensington Park Gardens. For some twenty years William Crookes, Charles Tidy and William

Odling (succeeded by James Dewar in 1894) had been paid by the consortium of London's private water companies to assess water purity and quality on a daily basis to challenge Edward Frankland's official claims that water quality was often contaminated by sewage. Because Crookes's home laboratory was cluttered by filtration bottles from his radiant spectroscopy research and Dewar's Royal Institution laboratory was filled with low-temperature apparatus, the two men fitted up a water analysis laboratory in Colville Road, a quarter of a mile from the homes of William and Henry Crookes. The premises appear to have been used by Dewar and Crookes for private consultancy work and it was registered in street directories as Crookes Analytical Chemists. Henry was placed in charge and the laboratory was continued up until the First World War—even after the new London Metropolitan Water Board abolished the advisory service of Crookes and Dewar in 1905 and set up its own laboratory and team of analysts. Here the Board analysed water quality on an hourly basis, something that William and Henry Crookes and Dewar would never have been able to do. Henry's analytical work included the bacteriological examination of water supplies outside London.

By the 1890s, when Henry Crookes entered into water analysis, bacteriological assay of water samples had become accepted as part of the scrutiny of water samples. It was Henry who began to experiment with mixtures of potassium permanganate and acetic acid as a bactericide, thus inaugurating the process of water purification that culminated in those of chlorine and ozone in the twentieth century. It was during his search for a superior bactericide that Henry began to examine the bactericidal effects of colloidal silver.

Henry was now beginning to make a name for himself and his father clearly aimed to get him elected to the Royal Society. He joined the Chemical Society in 1902. Henry had visited the West Indies in 1902 and photographed a volcanic eruption of La Soufrière on St Vincent in May of that year. His father proudly arranged for these photographs to be given to the West Indian Volcanoes Committee and displayed at a Royal Society Conversazione in Burlington House on 18 June 1902. Sir William by this date was actively researching radioactivity and Henry was able to use some of his father's radium to assess its anti-bacterial properties. At another Royal Society Ladies Night Conversazione in June 1903, Henry was able to exhibit photographs of plate cultures showing the deadly properties of radium. Of course, while very efficient as a bactericide, the cost of radium was too great to make this finding of interest to water companies.

In June 1911 Henry produced photographs and cultures for the Royal Society of *Bacillus phosphorescens*, demonstrating how the Petri dish cultures were destroyed by metals such as silver, mercury, antimony, thallium, copper, gold and platinum. (Such metals interfere with bacterial enzymes responsible for growth.) It was clear that he had found a promising line of research on germicides, a line that had obvious potential in the realms of medicine and commerce. What led Henry Crookes to collosols? The key factor was his business as a water analyst since this frequently involved him in sampling spa waters for bacterial contamination. Around 1903 he (and others) noted that springs famous for their health-giving properties frequently contained metallic particles in suspension engaged in Brownian motion—itself a molecular phenomenon then engaging the attention of Einstein, Zsigmondy and others. Perhaps, then, it followed that the health-promoting quality of these spa waters could be enhanced further by increasing the number of minute metallic particles in suspension. In 1898 Georg Bredig (1868–1942), then assisting Wilhelm Ostwald at the University of Leipzig, had found that he could prepare metallic colloids by passing a large electric current between two wires of a metal immersed in water or a suitable solvent. The metal was vaporized and condensed in the solvent as ultra-fine particles. Bredig likened these colloidal solutions or sols to "inorganic ferments" because of their catalytic effects. Bredig's work inspired The Svedberg (1884–1971) to use colloids for quantitative physiochemical studies. Henry Crookes, on the other hand, was inspired to use Bredig's sols as germicidal agents analogous to radium.

The immediate problem, however, was the fact that such sols were unstable, rapidly breaking down in the presence of electrolytes such as salt. Unless stabilized, their use as bactericides in the human body would be impossible. Although physical chemists soon cracked this problem, the resultant metallic sols did not satisfy Crookes's further need for them to be non-caustic and non-irritant on the skin. He apparently solved this problem after thousands of experiments in 1912, calling the resultant harmless metal colloids, "collosols". With the aid of medical co-workers he found that these could be applied safely as lotions, injections or even orally—provided they were sufficiently diluted.

Although the secret of the preparation was not revealed in patents, the trick presumably lay in preparing the more viscous lyophilic, rather than lyophobic, sols of silver and mercury since these are relatively unaffected by small amounts of electrolyte. Crookes speculated that bacteria were electrocuted by absorbing the negative charges of the metallic particles. Externally, bacteria were commonly killed within four to six minutes of application; colloidal silver remained unharmed by stomach acid and did not discolour the skin (argyria) as Barnes's Argyrol was prone to do. Dilution in not quite homeopathic proportions prevented them from poisoning human hosts.

Accordingly, Henry, together with a scientifically minded solicitor named Leslie Stroud, filed several patents for methods of making and stabilizing metal colloids between 1912 and 1913. They also contracted with a company promoter, Ernest Govett, to finance the flotation of a company called "Crookes Collosols" to begin their manufacture and sale. Govett agreed to buy the patents for £18,000 and £45,000 in £1 shares. The company planned to have a capital of £100,000, with a working capital of £22,000.

Unfortunately, the work was carried out at the Water Laboratory in Colville Road owned by William Crookes and James Dewar. The latter, a notoriously difficult employer, became convinced that he and the Royal Institution (which he directed) should receive financial benefit from the development of collosols, about which he had not been consulted. To make matters worse, Dewar accused Henry of appropriating some of his own observations (unspecified) and Sir William of breach of contract. In November 1912 he took out an injunction against father and son, as well as Stroud, and demanded Sir William's resignation as Secretary of the Royal Institution. In retaliation, the hitherto quiet and unassuming Henry sued Dewar for libel. Sir William, poised to be elected President of the Royal Society could not afford to be associated with scandal of any kind. Accordingly, he publicly dissociated himself from Henry's actions, claiming that Henry had never consulted him on his business affairs. Dewar's action in Chancery seems to have been settled by Sir William's action and by Henry agreeing to make any further development of collosols at his own home laboratory in Ladbroke Grove.

Although this absolved Sir William Crookes of commercial taint and allowed his election as President of the Royal Society in 1913, his action was undoubtedly specious. He had been instrumental in introducing Henry to Dr Henry Plimmer, a pathologist at St Mary's Hospital, who had encouraged Henry to develop collosols as germicides. Moreover, there can be no doubt that William helped Henry financially and that he invested in the Crookes Collosols Company. Sir William must have realized that his son was bound to trade on his father's association with the firm, even if it was restricted to investors. Indeed, to this day, Crookes Healthcare, now a subsidiary of Reckitt Benckiser, asserts that both William and Henry Crookes founded the firm in 1912. Significantly, throughout the 1920s and 1930s, the company used a drawing of Sir William, not Henry, as a trademark in advertisements.

Crookes Collosols received a second blow when Govett suddenly broke his promise to finance the company. His reasons are unclear, though it could well be because he decided it would be foolish to launch a new company when war was threatened. Stroud (on Henry Crookes's behalf) immediately sued Govett in the High Court for breach of contract, claiming £30,000 in damages. Govett did not try to deny that there was firm evidence of a contract having been made and offered no defence. Stroud's QC (J. B. Mathews) argued that Henry Crookes's invention had been damaged commercially because the delay in launching the company had allowed competitors to enter the market—a point confirmed to the magistrate by the manufacturing chemists, Messrs William Oppenheimer. In August 1914 the High Court found in favour of the plaintiffs, but limited damages to only £5,000, a sum that was probably largely swallowed up in legal fees. It is perhaps significant that Henry Crookes died intestate, leaving his widow only £300.

Despite these legal problems, Crookes Collosols began trading and selling "collosol argentums" in the early months of 1915. According to Fournier D'Albe, the first biographer of Sir William Crookes, the Dewar and Govett affairs led to Henry's bankruptcy, a calamity that his father refused to save him from. The bankruptcy was angrily denied by Madalina Crookes after the biography appeared in 1923. In fact, there is no evidence that Henry was ever bankrupted; probably D'Albe confused Henry with his brother Walter, whose practice as a solicitor had collapsed in 1911. Henry's wayward career ended tragically with his early death in

August 1915 from malaria and before he saw the fruits of his collosol research. He was buried in the family grave at Brompton Cemetery. Just a year before his death he had read a paper on the therapeutic benefits of collosol treatment to the Royal Society of Medicine. The paper's appearance in *Chemical News* suggests that any difference between father and son over Dewar and the Royal Institution had disappeared. Crookes even allowed his son a footnote stating that the collosols were manufactured by Crookes Collosols Ltd and were obtainable from the company's laboratory at 109 Ladbroke Grove, Henry's home address.

After Henry's death the silver colloid patents were bought by the dermatologist Malcolm Morris (1849–1924), who served as the company's medical director and moved production to premises at 50 Elgin Crescent, close to Henry Crookes's former home. The company was renamed Crookes Colloids in 1918 when it operated from premises near Tottenham Court Road. A year later it was rebranded as British Colloids Ltd. In 1924, following claims in the *British Medical Journal* that the company's products were not true colloids, Sir William Pope, professor of chemistry at Cambridge, was drafted to make a rigorous defence and subsequently made the company's chemical consultant. The company built new premises at Park Royal in the 1930s, becoming Crookes Laboratories and Crookes Healthcare in the 1960s. In 1971 the successful pharmaceutical business was acquired by Boots which continued to use "Crookes" as a label for the marketing of brands such as Nurofen, Strepsils, Optrex and skincare creams. In 2006, however, Boots sold the Crookes subsidiary to Reckitt Benckiser—itself a chemical cleaning products company dating back to the 1820s. Crookes Healthcare products currently provide Reckitt Benckiser with a new platform for over-the-counter health products. Meanwhile, silver collosols (marketed as "Silvagen") continue as important alternatives to antibiotics in the war against dangerous bacteria.

CHAPTER 24

A Life of Magic Chemistry

One of the dreams of sixteenth- and seventeenth-century practical chemists was that of the *Alkahest* or *Ignis-Aqua*, a universal solvent which, according to distinguished authorities such as Paracelsus, Joan Baptista van Helmont and Johann Glauber, had the extraordinary power of converting any substance into the primary matter that was supposed to be the root of all things. Access to it would be an important stage to the goal of alchemical transmutation. Paradoxically, this remarkable solvent could be contained in a glass bottle. Glauber curiously identified it as *sal mirabile* (sodium sulfate) but the sceptical Johann Kunckel dismissed the idea of an alkahest in 1716 with the pun *alles lügen est* [all lies].

In the mid-twentieth century, after chemistry had undergone a massive conceptual reshaping with the electronic theory of valence, solvents again became a subject of controversy. According to the ionic theory, acids were hydrogen-containing compounds that produced hydrogen ions in water. In 1923, taking into account solutions in non-aqueous media, Martin Lowry and Johannes Brønsted independently recommended that acids should be redefined as proton donors, the acid and base (proton acceptor) being seen as a conjugate to one another. The American chemist G. N. Lewis generalized this further in 1938 by defining acids as any substance (*e.g.* BCl_3) that could accept an electron pair. Acid strength, hitherto measured in terms of hydrogen ion activity in

The Case of the Poisonous Socks: Tales from Chemistry
By William H. Brock
© William H. Brock, 2011
Published by the Royal Society of Chemistry, www.rsc.org

aqueous solution, thereupon became a measure of the extent to which protons were released or electron pairs gained. This relating of acidity to the degree of transformation of a base into its conjugate acid proved particularly helpful in the development of post-war organic chemistry. In the 1960s, building on an earlier idea of James Conant's, Ronald Gillespie identified acids stronger than 100% sulfuric acids as "superacids". By combining strong acids, such as perchloric and trifluoromethanesulfonic acids, with Lewis acidic metal fluorides of arsenic and antimony, a range of these superacids was prepared. These included the so-called "magic acid"[®], a conjugate of fluorosulfuric acid with antimony pentafluoride, $FSO_3H–SbF_3$. The name was registered commercially by George Olah and Ned Arnett in 1962.

Such superacid systems, which can be 10^{16} time stronger than 100% sulfuric acid, had extremely low nucleophilicity. Consequently, when used as low temperature solvents, they could "freeze" the unstable ionic intermediates (carbocations) that many organic chemists believed, but could not prove, took part in reaction mechanisms. Olah had found an *alkahest* that would prove the existence of the primary matter of twentieth-century organic chemists, the carbocations, or unstable intermediates.

The existence of such carbocations (a term coined by Olah in 1972), which involved positing that carbon could have valence states of three and five as well as the usual four, proved extremely controversial—nowhere more so than in the Wagner–Meerwein rearrangements of camphene hydrochloride and other terpenes. Christopher Wilson and his mentor Christopher Ingold, as well as Saul Winstein and Olah, all favoured the fleeting existence of a non-classical ion to explain why only one of the two possible isomeric forms of bornyl chloride (the *iso* form) were formed in the rearrangement. Fierce critics, like Herbert C. Brown (1912–2004), who refused to accept electron delocalization in the dispersed phase, opted for a classical (quadrivalent carbon) explanation in terms of steric effects. The scene was set for some forty years of bitter controversy. At stake were classical valence theory *versus* the applicability of quantum chemistry and new experimental and instrumental techniques to the study of reaction mechanisms.

Olah had long been committed to the idea of charged hydrocarbon ions through his application of Freidel–Craft reactions to the study of hydrocarbons. By dissolving alkyl halides in "magic acid" at low temperatures he was able to stabilize the lifetimes of trivalent carbocations (carbenium ions) so that they could be identified by the application of nuclear magnetic resonance (NMR) and electron spectroscopy. In 1969 Olah showed that the penta-coordinated carbon in 2-norbornyl cation could be identified by NMR, but Brown challenged the result. Winstein died in the same year and Olah found himself defending the non-classical bridge until 1983, by which time he and others had produced over-whelming experimental evidence in favour of the fleeting existence of non-classical ions as intermediates in organic reactions. A Hungarian colleague of Olah aptly observed the importance of having a few good enemies in science to stimulate ideas and experiments! For these contributions to carbocation chemistry Olah was awarded the Nobel Prize for Chemistry in 1994.

Born in Hungary in 1927, Olah received a Germanic education at the Budapest Technical University where his post-doctoral work was on natural products chemistry. In the 1950s, when he wrote the first textbook in Hungarian on physical organic chemistry, his interest turned to fluorinated carbohydrates. In order to under-stand the mechanisms involved in their chemistry he was led to prepare and study superacid systems. The isolation of Hungary under the post-war Soviet regime was personally frustrating and he seized the chance that the 1956 Hungarian uprising provided to escape and forge a new life in Canada. Seven years of industrial work with Dow Chemicals saw him establish an international reputation as a hydrocarbon chemist specializing in petroleum chemistry. In 1965 he returned to academia, first at Cleveland's Case Western University and, from 1976, at his own privately funded Loker Hydrocarbon Research Institute at the University of Southern California in Los Angeles. He observed wryly that it is not enough to be Hungarian to succeed; you must also be talented and leave the country.

Part 4: Women Chemists

If the general public finds it difficult to identify a famous male chemist, with the possible exception of Madame Curie, women chemists are even more "hidden from history". There has been much debate whether their invisibility in the history of chemistry is because women chemists never did anything of importance or, as is more probable, that they have never been sufficiently acknowledged. There is also an assumption that women only began to take up chemical careers after the mid-twentieth century. But as the first story reveals, women are known to have practised alchemy and pharmacy well before the seventeenth century. It is true, however, that the development of the physical sciences from the seventeenth century onwards tended to be a male preserve. This limited women to secondary roles such as that provided by Madame Lavoisier in recording experiments for her more famous husband and drawing diagrams for him. Roles in popularization were also possible for determined women such as Mrs Jane Marcet, whose *Conversations of Chemistry* (1807) were inspired by listening to lectures by Humphry Davy and which persuaded several men to take up chemistry. Our second story underlines the supportive roles that spouses play in sustaining and encouraging their husbands' careers whether or not they themselves have any chemical knowledge.

Despite these qualifications it is true that until higher education became available to women in the last decades of the nineteenth century it was virtually impossible for women to work in laboratories or to pursue an industrial career. The story of Ida Freund shows how women's exclusion from chemistry was overcome by the establishment of their own teaching laboratories. Such career moves were viewed with suspicion by male chemists like Henry Armstrong. When he visited the United States in 1903 on a fact-finding mission on American science education he thoroughly disliked what he saw of women's advancement in the States. This was despite the fact that he had admitted a few exceptional women into his own laboratory such as Grace Heath (later a science mistress at North London Collegiate School) and Louise E. Walter who became a school inspector. But he drew the line at sharing laboratory research space for women by repeatedly campaigning against the admission of women to the Chemical Society. In 1904 the admission of women was decisively defeated by legal opinion by which time Armstrong was willing to offer them secondary status as "lady subscribers" but otherwise muddling the issue of their membership with suffragette agitations. It was not until 1920 that women were allowed to become Fellows of the Chemical Society beginning with the admission of Hilda Ingold's Imperial College supervisor, Martha Whiteley, and not until 1940 that women were admitted to the Royal Society.

Like nurses in hospitals it was only too easy for graduate women chemists to have their potential research careers scuppered by marriage. For example, Annie Sedgwick's career finished when she married the physical chemist James Walker. Other wives such as Marian Partington and Peggy Fowden were able to continue their chemistry as schoolteachers. Chemical partnerships such as those of Pierre and Marie Curie, Robert and Gertrude Robinson, Christopher and Edith Ingold, or Louis and Mary Fieser in America were exceptionable before the 1950s, though now much more common. Where chemical partnerships were not formed, some husbands, like Thomas Lonsdale, were willing to encourage their talented wives in a research career. It was in this way that Kathleen Lonsdale (1903–1971) maintained her brilliant career in crystallography. Such partnerships, in which husband and wives pursue independent research careers rather than collaborating have become more and more possible since the late twentieth century. Yet it still remains a problem of how to encourage young women to take up chemistry as a career.

CHAPTER 25

Women in Alchemy and Chemistry

The possibility of manipulating matter into substances of commercial value such as silver and gold on the one hand, or of physically uplifting value, such as an elixir of life on the other, was known as alchemia or alchemy. It was one of the roots of chemistry and biochemistry as well as chemotherapy. Alchemists derived much of their apparatus and manipulative techniques from the equipment used and developed by artisans, metallurgists and pharmacists whose heating, cooking, subliming and distilling techniques became grist to the mill of chemistry as it emerged in the sixteenth century. A striking example of this continuity between alchemy and chemistry is found in the water bath as well as in distillation apparatus generally (Figure 25.1).

The development of the water bath, or *bain marie* as it is still known in Europe, is attributed by Zosimus, the greatest authority among Hellenistic alchemists, to a female alchemist named Mary the Jewess. We know nothing about where she lived, or when she was active, though from her references to God as having directly revealed the secrets of nature to her and that they were not to be passed on to those who were not of "the race of Abraham", her Jewishness is confirmed. Zosimus quoted extensively from one of her manuscripts entitled *On Furnaces and Apparatus* and it is from this that we know she constructed and used various devices made from glass, clay and metals to sublimate and distil materials like mercury, which she described as a "deadly poison, since it dissolves

The Case of the Poisonous Socks: Tales from Chemistry
By William H. Brock
© William H. Brock, 2011
Published by the Royal Society of Chemistry, www.rsc.org

Figure 25.1 A medieval water bath developed from the original form of *bain Marie* attributed to Mary the Jewess in the third century AD. (Wellcome Images)

gold, and the most injurious of metals". Among her appliances, no doubt derived from cookery, was a double vessel, the outer one of which was filled with water and heated while the inner vessel contained the materials to be treated by the gentle warmth. It seems quite probable that, like Liebig's condenser, the water bath had been in use for centuries before and that it was the fame and prestige of Marie's writings that led to it being forever associated with her name. Whatever the truth of the association, her book on apparatus, which also included accounts of various forms of still, influenced later Arabic alchemists and through them, practical alchemy in Europe. It is astonishing to realize that so many of the basic laboratory techniques that chemists still use in the laboratory owe their origin to alchemists working in the third century AD and that their elaboration and perfection was probably due to a woman practitioner (Figure 25.2).

Once alchemical practices had entered the Latin West it seems to have been quite common for women to claim expertise in the art. Although they do not appear as writers themselves, manuscript and book illustrations frequently portray men and women sharing chemical manipulations in a workshop. Such partnerships could be dangerous, as the case of young Anna Maria Zieglerin (c. 1550–1575) demonstrates. Her case also helps underline the fact that alchemy was applied chemistry and embraced far more than gold-making. By the sixteenth century it was extremely common for European rulers and rich land-owners to employ alchemists to help enrich them by exploiting the minerals and metals on their estates.

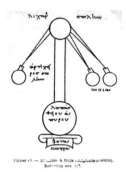

Figure 25.2   Greek distillation apparatus of the type described by Mary the Jewess and other Greek alchemists in the third century AD. (Wellcome Images)

Alchemists are too easily written off by later writers as frauds and deliberate deceivers; it now seems far more likely that they were often condemned for failing to adhere to an employment contract.

In 1574 Julius, the Duke of Braunschweig-Wolfenbüttel, took criminal proceedings against a number of alchemists in his employ, including Heinrich Schombach and his wife Anna Zieglerin. They were accused not only of attempting to poison the Duke's wife and stealing private documents, but of failing to complete alchemical projects they had been paid for. According to the strange evidence against her (mostly drawn from her husband), Anna Zieglerin had been born prematurely by Caesarean section, nurtured as a baby in a woman's flayed skin, and supposed herself as pure as the Virgin Mary. At the age of fourteen she was raped by a rejected suitor and bore a son which she drowned. Some years later she married the cross-eyed court jester, Heinrich Schombach. In 1571 Anna and her husband, for whom she showed little respect, together with another alchemist named Phillip Sömmering, were contracted by Duke Julius to work on various alchemical and mining projects. All went well initially as, working in her own laboratory, Anna sent the Duke regular reports on the practicalities of preparing the philosopher's stone which, in turn, she promised would replicate precious metals and jewels. The stone would also enable the production of exotic fruits and effective medicines, as well as healthy children. She claimed to have been taught alchemy by Carl von Oettingen, the fictitious son of Paracelsus, who had wanted to marry her!

As a female consultant, Anna's captivation of Duke Julius's attention must have seemed a threat, particularly to his wife and the Duke's other male advisers, who ganged up against the alchemical trio to oust them from the court. Anna's end was gruesome. After confessing to their crimes Anna and the others were flayed alive and then burned to death strapped upon an iron stool. As the American historian of alchemy Tara Nummedal has concluded in telling Anna's story: alchemy was a "high risk, high reward game in the early modern Holy Roman Empire". But it was a game played by both sexes.

CHAPTER 26

Teaching Chemistry to Women

A Woman's college! Maddest folly going!
What can girls learn within its walls worth knowing?

One answer to W. S. Gilbert's satirical query in the comic opera *Princess Ida* (1884) was, of course, chemistry. And Newnham College, an all-female college established at the University of Cambridge in 1871, provided some of the best chemical education of any of Britain's higher educational institutions under the instruction of a real-life "Ida", the Anglo-Austrian chemist Ida Freund (1863–1914) (Figure 26.1).

Freund made an unusual figure in Cambridge. She had lost a leg in a cycling accident as a child in Vienna and subsequently walked around the laboratory on a wooden leg. She continued to cycle all her life, using an adapted vehicle whose chain she revolved from the handlebars. She was orphaned early in life and brought up by grandparents and other relations in Vienna where she had a good secondary education. Her uncle, the professional violinist, Ludwig Strauss, brought her to London in 1881 when he was on a concert tour with the Joachim string quartet and she decided to complete her education in England. A year later she matriculated at Girton College, Cambridge's original women's college which had been founded in 1869. There she took the Natural Sciences Tripos with chemistry as her specialism for the final year. Although awarded first class honours, like other female graduates, the degree was

The Case of the Poisonous Socks: Tales from Chemistry
By William H. Brock
© William H. Brock, 2011
Published by the Royal Society of Chemistry, www.rsc.org

Figure 26.1 Ida Freund (1863–1914), an Anglo-Austrian physical chemist, she pioneered the teaching of chemistry to women at Newnham College, Cambridge. (Newnham College)

denied her. After a year's teaching at the Cambridge Training College for Women (later named Hughes Hall), in 1887 she was appointed a demonstrator at Newnham and this was made into a chemistry lectureship a few years later. She was the first woman lecturer in chemistry at a British institution of higher education.

Newnham College had its own laboratory since men and women were not allowed to work together until 1910. Freund was given special responsibility for running this laboratory and although she conducted some research in physical chemistry, she ran it very much as a teaching workshop. Like her contemporary, Henry Armstrong, she deeply believed that chemistry was best learned at the laboratory bench. She went further than Armstrong, however, in insisting that her students should read original papers and test the validity of published work. She was particular about accuracy in weighing and measuring, and in testing the basic assumptions of chemistry like the law of constant composition. She summed up her approach to teaching, in which she was quite original, in *The Study of Chemical Composition* (1904), a monograph which combined formidable historical scholarship with deep analysis of scientific reasoning.

Not surprisingly in view of her work at the Women's Training College and at Newnham, Ida Freund was concerned to see improvements in science teaching in girls' schools. The chemist Edward Frankland, together with his colleagues T. H. Huxley and Frederick Guthrie, had introduced summer schools for male science teachers at the Royal College of Science in South Kensington in the 1870s. Frankland's experimentally-based course, *How to Teach Chemistry*, had been published by one of his pupils in 1872. Accordingly, in 1897 Freund adapted Frankland's model to improve the standard of laboratory teaching in girls' schools by holding summer vacation classes for women chemistry teachers at Newnham. Once again the course was concerned with demonstrating the fundamental principles of chemistry. Her experience on this programme was summed up in the book *The Experimental Basis of Chemistry* which was unfortunately not completed before her early death in 1914. It was carefully edited by her pupil M. B. Thomas and finally published in 1920.

Mary Beatrice Thomas (1873–1954), who graduated from Newnham in 1898, held an equivalent position to Freund's at Girton College, but was keener on research. Nevertheless, as Director of Science Studies at Girton, where she is said to have trained some 300 women scientists between 1906 and 1935, she kept faith with Freund's insistence on an understanding of the fundamental laws of chemistry. Cambridge was one of the last bastions of resistance to women's graduation and it was not until 1948 that women graduates were treated equally to men. However, the opening of the main university laboratory to women in 1910 meant that women's degrees would have to be eventually recognized. Newnham closed its laboratory immediately after Freund's death, but Girton delayed closure until 1935 when Thomas retired. Freund and Thomas between them had ensured that women chemists educated at Cambridge were equipped to be skilful science teachers, chemical librarians and laboratory assistants and that some, like Rosalind Franklin (1920–1958) would be able to produce revolutionary research.

CHAPTER 27

Musical Affinities

Medical practitioners have long had a sideline as novelists. Some, like Arthur Conan Doyle and A. J. Cronin, even found a literary life more stimulating and profitable than medicine. Chemist novelists have been a rarer species and largely confined to organic chemists. Edward Frankland's grandson, Edward Percy Frankland (1884–1958), abandoned chemistry in the 1920s to become a farmer and a prolific writer of romances for the old circulating libraries, as did Alfred W. Stewart (1880–1947), who found wider fame as the writer of detective fiction under the *nom de plume* of J. J. Connington. More recently, Carl Djerassi, after a lifetime of steroid research, has enjoyed considerable success as a writer of science-based novels such as *Cantor's Dilemma*.

Not all literary chemists have been successful. In 2001 the American nonagenarian silicone chemist, Eugene Rochow (1909–2002), produced a novel about real chemists with the assistance of a younger German chemist, Eduard Krahé. *The Holland Sisters* fell dangerously between the twin stools of fiction and straightforward narrative that provides the social and intellectual context in which real historical characters move. This was a "romantic historical novel" that did not come off. Rochow was stymied by the fact that none of the sisters' diaries or letters has survived. Everything concerning the sisters' upbringing, how they lived and what they thought and did has to be conjured up from knowledge of (or rather a potted history of) the educational, political and social events of the times.

The Case of the Poisonous Socks: Tales from Chemistry
By William H. Brock
© William H. Brock, 2011
Published by the Royal Society of Chemistry, www.rsc.org

The imagined conversations are stilted and unnatural, reminding the historian of the improving didactic conversations of early nineteenth-century chemistry popularizers such as Samuel Parkes and Mrs Alexander Marcet. British readers particularly found the American cultural bias in telling the story rather intrusive. Indeed, in writing for an American readership, there was far too much irrelevant padding concerning two world wars.

Despite this negative reaction, the story of the Holland sisters was worth telling. They were surely unique in the annals of science in marrying three brilliant chemists. The three musically talented sisters of the story, Mina (1865–?), Lily (1867–1949) and Kathleen (1879–1960), were the daughters of William Holland, a prosperous brick and tile manufacturer of Bridgwater in Somerset. Like many Unitarians, Holland participated actively in local government and was twice elected Mayor of Bridgwater. Their mother, Florence Du Val, was an artist's daughter from Manchester. Her sister had married into a family named Kipping and her son, Frederic Stanley Kipping (1863–1941), who became the famous silicon chemist, was a frequent visitor to his aunt's and cousins' home in Bridgwater. A mutual passion for tennis and music led Lily to fall in love with him when she was still a schoolgirl (Figure 27.1).

While working as a research student in Munich, Kipping met William Henry Perkin (1860–1929), a fine pianist and son of the great dyestuffs chemist. They became firm friends and immortally linked by the best-selling organic chemistry textbook that they first produced in 1894. Both Kipping and Perkin found posts at the University of Manchester and it was there, during a family visit from the Hollands, that Perkin and Mina Holland fell in love, music again being a mutual passion. Following the convention that an eldest daughter married first, Perkin and Mina started their lifelong partnership in 1887, settling first in Manchester and, from 1912, in Oxford, where Perkin created the school of organic chemistry that dominated twentieth-century English research. Lily married Kipping in 1888, when he was elected to a chair at Heriot-Watt College in Edinburgh. After a period in London, the Kippings finally settled at the University College at Nottingham, where Kipping established his international reputation as an organo-silicon chemist (Figure 27.2).

Finally, at another family gathering in Bridgwater, the youngest sister, Kathy, met Kipping's doctoral student, Arthur Lapworth

Figure 27.1   William Henry Perkin junior (1860–1929), the son of the dyestuffs chemist, Sir William Henry Perkin (1838–1907). As professor of chemistry at Oxford from 1892 he built up a great school of organic research. He was the brother-in-law of both Lapworth and Kipping. (Science & Technology Picture Library)

Figure 27.2   Arthur Lapworth (1872–1941), a pioneer of the electronic theory of valence. He was a brother-in-law of both W. H. Perkin Jr and F. S. Kipping. (Brock)

(1872–1941). Once again music was a common bond, Lapworth being a skilled violinist and Kathy a fine cellist. They were married in 1900 when Lapworth obtained a post at the School of Pharmacy in London. In 1912 they moved to Manchester when Lapworth succeeded Perkin as Professor of Chemistry. Unlike the Perkins' and Kippings' marriages, the Lapworths' marriage seems to have been the least successful of the partnerships, though it was kept alive by their mutual love of chamber music.

Henry Armstrong once described Perkin as "mad as several Mad Hatters in his devotion to research", and the remark could be applied equally to Kipping and Lapworth. The extraordinarily long hours they worked in their laboratories must have meant their wives saw very little of them except at weekends and for holidays. How then did they influence the success of these chemists?

In recent years there have been many books and articles on female scientists and the difficulties they have experienced in the male-dominated areas of teaching, writing and research. Although feminist historians have acknowledged the roles of wives, sisters and daughters in assisting husbands, brothers and fathers with their scientific research, *The Holland Sisters* was a pioneering attempt to portray the domestic support that has necessarily sustained male academic careers. However, in the absence of documentary evidence the authors had difficulty in going beyond sketches of scenes of loving domestic anchorage in the Perkin, Kipping and Lapworth households. There was no implication that the sisters contributed directly to their husbands' research which, apart from Kipping's preparation of silicones, was surprisingly under-described. The reader would never guess, for example, that Lapworth made fundamental contributions to the electronic theory of valence. The authors did, however, stress that Mina Perkin was a great social asset for Perkin at Oxford. They also allowed Lily Kipping (the most chemically informed of the three sisters) to help Kipping to prepare later editions of the famous organic chemistry text. The least domestic of the three sisters was the childless Kathleen Lapworth, who (like Hilda Ingold at University College London) administered the Manchester chemistry department until her husband's retirement in 1935.

Perhaps if Rochow and Krahé had used a first person narrative, choosing Lily Holland (Kipping's wife) as the narrator, they could have produced a satisfying novel. In the event, a third person format with an omniscient narrator proves unsatisfactory. Nevertheless, the book is well worth reading as a chemical curiosity. Perhaps another chemist will be persuaded by it to try their hand at a novel about a chemical family? One obvious candidate is the Frankland dynasty most of whose members kept diaries and journals and whose lives were dominated by feuds and rivalries, *and* writing novels.

CHAPTER 28

Edith Hilda Usherwood (1898–1988) and the Ingold Partnership

Despite a wave of interest in the careers and roles of women scientists in the late nineteenth and early twentieth centuries and the forty-year struggle by women chemists for admission to Fellowship of the Chemical Society between 1880 and 1920, little has been written about husband–wife partnerships. Here we look at the career of Edith Hilda Usherwood, better known as Mrs Hilda Ingold or Lady Ingold. The devoted wife of the physical organic chemist Sir Christopher Kelk Ingold (1893–1970), who always called her Hilda (not Edith), she died in 1988 without any national recognition. This was in contrast to Gertrude Robinson (1886–1954), the wife of Christopher Ingold's great rival, Sir Robert Robinson (1886–1975). Yet like Gertrude, Hilda had been a chemist of considerable originality in her own right, her husband's collaborator, and a significant helper in his becoming a chemist of international renown. Such overshadowing of wives by their husbands has been called the "Mathilda effect". This plays upon what sociologists had previously identified as the "Matthew effect" whereby one or more collaborators in a piece of research can be ignored in favour of the first-named person in a list of authors. Similarly, wives in a husband–wife partnership may find themselves playing second fiddle while the husband gains the credit (Figure 28.1).

The Case of the Poisonous Socks: Tales from Chemistry
By William H. Brock
© William H. Brock, 2011
Published by the Royal Society of Chemistry, www.rsc.org

Figure 28.1   Edith Hilda Usherwood (1898–1988), an organic chemist who married and collaborated with Christopher Kelk Ingold on the electronic theory of valence. This photograph was taken in 1940 when University College London was evacuated to Aberystwyth. (UCL Chemistry)

Hilda Usherwood was born at Catford, London, on 21 May 1898, the daughter of a mechanical engineer, Thomas S. Usherwood, who had been trained at the City & Guilds' Central College, South Kensington, where his talents had been spotted by the chemist Henry Armstrong. When Christ's Hospital School moved from London to Horsham in West Sussex in 1908, through Armstrong's persuasion, the school opened a Science, Manual and Art Department. The school was to be a model of heuristic teaching. Three of Armstrong's pupils and friends were appointed as teachers, including Usherwood who was selected to lead the Manual or Engineering School modelled on the one the great headmaster F. W. Sanderson had developed at Oundle. The Usherwood family consequently moved to Sussex, and from 1906 to 1909 Hilda attended the Girls' Grammar School in Lewisham. Following a further two years of private education in Horsham, in 1912 she won a scholarship to North London Collegiate School, whose headmistress was the mathematician and pioneer of higher education for women, Dr Sophie Bryant. Although quite elderly by 1912, Bryant was an inspiration for any girl with mathematical or scientific aptitude. The scientific atmosphere of North London Collegiate, coupled with Usherwood's practical engineering and heuristic background, may well explain Hilda's bent towards physical organic chemistry and her receptivity towards the new quantum mechanics in the mid-1920s. Interestingly, Hilda's chemistry teacher at North London Collegiate was Grace Heath, another former pupil of Henry Armstrong's at the Central Institution during the late 1880s. Usherwood therefore almost certainly received a heuristic, manually based chemical education and the evidence from her published papers suggests that she was a deft experimentalist. Given C. K. Ingold's bent towards physics while a student at Hartley College in Southampton, the later marriage was to be a coupling of minds interested in explaining organic reactions physically or, as they began to say, mechanistically.

Following the entrance scholarship in 1912, Hilda was Clothworkers' Scholar for 1915–1916 when (in 1916) she won an entrance scholarship to Royal Holloway College, a pioneering women's college founded at Egham in Surrey in 1886 by Thomas Holloway from the fortune he had made from selling pills. It became part of the University of London in 1900.

There is a College down in Surrey, south west of London Town,
The best place in all England, where they wear the cap and gown.
For its emblem is a rose and mid roses fair it grows,
Set in leafy summer foliage, itself a red, red rose.
(College song)

Hilda registered for the BSc in honours chemistry with subsidiary physics and entered upon a brilliant student career. Altogether she won ten prizes at Royal Holloway, including one for music, and graduated with first class honours in October 1920. Her academic activity was combined with an active social life of hockey playing, tennis, rowing, piano playing, the secretaryship of the Music Club and the presidency of the Women's Settlement at Southwark where the social work expected of all lady undergraduates was carried out. She remained a Fabian socialist all her life. This was a rather more impressive record than that of her later husband's at

Southampton! Indeed, Michael Dewar in an interview (though not in his published autobiography) made a catty remark about Ingold's failure to gain first class honours from Southampton not appreciating, perhaps, how difficult it was to do this from the small provincial college that was the origin of today's University of Southampton.

Chemistry teaching at Royal Holloway had begun in 1891 with Mary Robertson, who was succeeded two years later by Eleanor Field, one of Ida Freund's pupils at Newnham College. Field's most brilliant student was Martha Whiteley (1866–1956) from South Kensington High School. Whiteley was to become the first woman chemistry lecturer at Imperial College, as well as one of the first women Fellows of the London-based Chemical Society in 1920. Field's only assistant was Dr Mary Boyle, a former student, who remained the College's Assistant Lecturer and Demonstrator from 1903 until 1933.

Chemistry was taught in a detached building in the College grounds that had been designed by the Oxford chemist A. G. Vernon Harcourt in 1889. It contained a lecture theatre, a general laboratory for thirty women, balance room and two small research laboratories. When Lord Haldane inspected the College in 1909, however, there were only 19 chemistry students. Since the number of students dropped dramatically during the First World War, it can be surmised that during Hilda's period classes were probably as small as four. In other words, she would have had almost personal attention from Thomas S. Moore, the Oxford-educated Professor who had succeeded Field in 1914 following the brief succession of George Barger (1913–1914) who had abruptly left to join the Medical Research Committee at the beginning of the war.

Moore had done some work on partition coefficients and had speculated about what was later called the hydrogen bond. (He is best known to historians as the co-author of a history of the Chemical Society published in 1947.) During the war, Moore worked on the electrolytic oxidation of ethylene to acetylene and introduced a scheme whereby third-year women carried out a research project. The files do not reveal what this was, but it clearly gained Hilda the College's Driver Prize. This, together with her first class honours, was more than sufficient to gain her a postgraduate place at her father's old college, which had since become a constituent of Imperial College at South Kensington. Here she was to work under Martha Whiteley who by then had become the chemistry department's first Assistant Professor working alongside Jocylyn Field Thorpe. By 1920 the department had some thirty members of staff and research workers, most of whom were working in organic chemistry; it was one of the largest research schools in the UK. Her financial support came from a grant awarded her by the Department of Scientific and Industrial Research.

We should recall that the doctoral degree (the PhD) had only been introduced into British universities in 1919 and was considered solely as research training. Consequently, it was easily fulfilled in two years or less, as Usherwood did. Given Whiteley's interest in the tautomerism of oximes, and the interest of the whole Imperial College group in tautomerism, not surprisingly this became the subject of Usherwood's short thesis: *Experiments on the Detection of Equilibria in Gaseous Tautomeric Substances. The Formation of Heterocyclic Rings Involving Reactions with the Nitroso and Nitro Groups in the Various Tautomeric Modifications.* Essentially it consisted of two research papers, together with an already published paper on the specific heats of hydrogen cyanide. Later, in 1925, Usherwood took the then still conventional DSc by submission of a portfolio of publications.

Carbon compounds containing the keto group ($CH_3CH=O$) with hydrogen atoms (or other groups) on the initial carbon atom of the structure are easily converted into their corresponding unsaturated alcohols (enols), the reaction being reversible.

$$H_3C.CH:O \rightarrow H_2C.CH.OH$$

keto-form ← enol-form

This rapid interconversion between two chemically distinct species of the same composition is an example of isomerism. One of the objections to the structural theory of organic chemistry that developed during the 1850s and 1860s was that some compounds appeared to display the properties of both ketones and alcohols. Thus, in some situations acetoacetic ester behaved as an alcohol (enol) ($CH_3.C(OH):CH.COOC_2H_5$), and at other times as a ketone ($CH_3CO.CH_2.COOC_2H_5$). The structural theory was saved by supposing that such molecules were isomeric; but unlike ordinary isomers, they easily changed from one form (enol) to the other (ketonic). In 1886 the German chemist Peter Laar named the phenomenon *tautomerism* (from the Greek *tauto*, the same, and *meros*, part) and the individual isomers, *tautomers*.

Tautomerism was of great interest among chemists at Imperial College. If the mechanism whereby double bonds were opened and closed, and labile protons (hydrogen atoms) moved from one site to another, could be understood, it held the possibility of explaining the course of chemical reactions more generally. Usherwood's doctoral thesis led to a joint paper with Whiteley on "The oxime of mesoxamide (isonitrosomalonamide) and some allied compounds" [*Journal of the Chemical Society*, 1923, **123**, 1069–1089]. This was actually the third of a series of papers by Whiteley on the formation of rings in tetra-substituted series of oximes, the latter being a carbon chain in which oxygen was replaced by a nitroso group, NOH. Oximes can exist in two forms according to whether carbon is attached to nitrogen by a double or single bond, $>C:NOH$ or $>C.NHO$, the latter being known as the *iso*oxime. Whiteley's aim was to see whether the reversibility of typical addition reactions was general, and how temperature and structure affected tautomeric equilibria. The experimental work, which fell to Hilda, required a considerable number of condensation reactions, combustions and syntheses to establish structures. The experimental details occupied thirteen pages of the printed paper.

Significantly, the thesis had involved a study of the equilibrium between hydrocyanic acid and its tautomer, hydrogen isocyanic acid, HNC. Hilda published her first solo paper on this "flickering" equilibrium in the *Journal of the Chemical Society* in 1922. Although others, such as the German physical chemist Conrad van Laar (1853–1929), had succeeded in demonstrating tautomerism in triads, no one had yet done it formally for dyads. (A triad contains three molecular units, while a dyad like hydrogen cyanide consists of only two.) Again, although chemical preparations, heats of combustion and refractivity studies all pointed to HCN as a nitrile and isonitrile mixture, this had never been formally demonstrated. Hilda did this, using the ingenious argument that since the shift of the equilibrium with temperature from HCN $\rightarrow$ HNC involved heat, q, which was much larger than the specific heat, a small shift of the equilibrium must add considerably to the amount of heat normally sufficient to raise the temperature by 1 °C. Consequently, the apparent specific heat would be greater than the true specific heat as calculated by the method of mixtures. This meant that the proportions of formonitrile (HCN) and carbylamine (HNC) present in the tautometric mixture could be calculated from the difference between the observed and calculated specific heats at constant volume over a temperature range. Using Kundt tubes to measure the velocity of sound (from which the ratio of specific heats at constant pressure and volume could be calculated) and living dangerously, she determined specific heats of HCN at various temperatures and concluded that gaseous HCN was a mixture of mainly formonitrile in equilibrium with about one per cent of carbylamine. Here was a splendid amalgamation of theoretical physical chemistry with highly dangerous chemical manipulation.

The work was, however, significant for another reason since it led her to collaborate with Jocylyn Field Thorpe's 28-year old assistant, Christopher Kelk Ingold, who was also engaged in tautomeric studies. Indeed, Hilda's first paper had thanked a "Dr Ingold" for "invaluable advice"—probably on how to handle gaseous prussic acid safely. There is anecdotal evidence that the two met when Ingold accidentally gassed himself in the laboratory with phosgene (carbon oxychloride) and Hilda came to his aid. However, this story is probably confuses an accident during Ingold's secret wartime work in Glasgow with the Cassell Cyanide Company when he was filling shells with "Jellite" (a mixture of hydrogen cyanide and arsenious chloride). When the delivery pipe suddenly became loose, Ingold had held his breath, reattached the pipe, and walked quickly upwind. It was for this kind of bravery that he had been awarded a British Empire Medal in 1920 though, characteristically, he never discussed this wartime work in later life.

Ingold had investigated heterogeneous equilibria while working for the Cassell Company and this had interested him in the variation of vapour pressure with temperature. This probably sufficiently accounts for Ingold being willing to collaborate with Usherwood in 1921 on the paper "Specific heats of gases with special reference to hydrogen" [*Journal of the Chemical Society*, 1922, **121**, 2286–2289]. In this theoretical paper they applied the principle of the equipartition of energy to diatomic gases and argued that factors for degrees of freedom of rotation and vibration must rise with temperature—and that this correlated with the observed specific heat curves of hydrogen, nitrogen and the halogen gases. They concluded that the constant volume specific heat curve (c_v against log T) fell into three sections, corresponding to the elements of the equipartition theorem. At low temperatures the heat component was due to translation only. At intermediate temperatures a quantum of rotation kicked in, followed later by a further rise caused by vibration quanta. From comparisons between the specific heat curves for hydrogen, nitrogen and chlorine, they were able to draw conclusions about the dissociation of diatomic gases at high temperatures.

Hilda's work on tautomerism and specific heats had three consequences for her. First, as we have seen, it led to her first solo paper on hydrogen cyanide coupled with an equally dangerous investigation of the chlorination of acetylene to form the inflammable explosive chloracetylene in 1924. If acetylene is considered as another dyad analogous to HCN, it should also display tautomerism. Using the octet symbolism of the American chemist G. N. Lewis in which bonds are portrayed as single, double and triple dots, and a mobile hydrogen as [H], Hilda argued for the analogy. The evidence was given to a meeting of the British Association in 1924.

$$[H] \; C ::: N \quad \rightarrow \quad C : N \; [H]$$

hydrogen cyanide   hydrogen *iso*cyanide

$$[H] \; C ::: CH \quad \rightarrow \quad C : CH \; [H]$$

acetylene   *iso*acetylene

Hilda's proof that such a tautomeric change occurred deployed two arguments. In the first place, she was able to determine the specific heat curve for acetylene using the same techniques she had used for hydrogen cyanide. The curve demonstrated that the unstable *iso*-form of acetylene was minimally present at ordinary temperatures, but it achieved appreciable concentrations at temperatures between 200 and 300 °C. The second argument involved oxidizing acetylene (a highly dangerous procedure) and showing that some of the products, the carbimides, could only have been formed from the *iso*-form of acetylene.

$$R.N:C + O \quad \rightarrow \quad R.N:C:O$$

*iso*nitriles   carbimides

$$R_2C:C + O \quad \rightarrow \quad R_2C:C: O$$

iso-acetylene   ketenes

The investigations not only showed that dyads like hydrogen cyanide and acetylene were tautomeric, but it also began to involve her in thoughts about reaction mechanisms generally.

The second consequence was that her work on the specific heat of hydrogen was challenged by J. R. Partington of Queen Mary College in 1925, the year after he and W. G. Shilling (1899–1955) had published a monograph, *The Specific Heats of Gases* (1924). In an article in the *Philosophical Magazine* in 1925, Partington suggested that Hilda's hydrogen results had been only approximate; more controversially, he also suggested that her experiments with HCN were not due to a thermal effect that accompanied an isomeric change, but were due to polymerization—a possibility she had ignored. Hilda replied, standing her ground; Partington stood his; and Hilda finally won her case by showing that Partington's evidence for association (polymerization) was valid only for a very small part of the temperature range she had investigated. As far as one can tell, she received (and needed) no help from C. K. Ingold in combating London's leading physical chemist.

The third and most important consequence was that the collaboration between Usherwood and Ingold led to marriage in July 1923. As one of Hilda's Royal Holloway College contemporaries, Kathleen Lathbury noted: "most of my generation of students did not expect to marry … because our opposite numbers had been killed … Teaching was the most obvious activity for which most of us studied". (Lathbury herself became a portrait painter after working for some years in the chemical industry.) Clearly, Hilda was one of the lucky First World War spinsters who married and had a family. The phenomenon of forced post-war spinsterdom does, however, raise an interesting question: did British women chemists became more visible and successful professionally after 1919 because many of them were unable to marry?

On Ingold's move to the University of Leeds in October 1924 as a professor of chemistry at the young age of 31, Hilda became an unpaid demonstrator, combining this with her social role as a professor's wife. She also became one of her husband's many partners and collaborators in the exciting and controversially productive period when Arthur Lapworth's and Robert Robinson's electronic theory was challenged and reinterpreted; when many cases of tautomerism were reinterpreted as mesomerism or resonance by means of the new quantum mechanics; and when the puzzle of electrophilic aromatic substitution began to be resolved.

The shift to the study of the mechanism of chemical reactions is clearly seen in the collaborative work that the Ingolds published between 1924 and 1928 in the field of what Martin Lowry had termed "protropy" (1923)—referring to the assumption that during tautomerism a hydrogen ion (proton) migrated. In 1927 the Ingolds renamed this mechanism as "cationotropy" to make it clear that the mechanism could be applied to any mobile cation. (Similarly, anionotropy referred to migratory anions like chlorine.)

The interest in prototropy arose when Thorpe needed a sample of methane-triacetic acid for his work on spiro-compounds in 1921. Since it was easily synthesized by traditional methods, Christopher Ingold devised a synthesis using the hydrolysis of ethyl-β hydroxygluarate and ethyl sodiocyanoacetate. He noticed that this Michael condensation was easily reversible and suggested to Hilda that she investigate this and similar three-carbon (triad) protrotropic systems. Hilda began with the simplest in 1923, namely keto–enol protropy in aldol condensation.

$$2CH_3.CHO \rightarrow CH_3.CH(OH).CH_2.CHO \text{ (50\% yield of aldol)}$$

$$\rightarrow CH_3.CH=CH.CHO \text{ (crotonaldehyde)} + H_2O$$

She adopted the principle that in any mobile triad system, $[H] - X - Y - Z$, there were two parts, $[H - X]$ and $Y = Z$, whose interunion produces tautomeric change *via* a mobile proton. By the dexterous alteration of the experimental conditions, Hilda was able to show how the production of an aldehyde or a ketone could be maximized.

In her first solo paper using her married name in 1924, Hilda examined additive reactions, particularly cyano–imino reactions, in relation to the stability of carbon chains. In her final solo paper, she investigated Lewis polar conditions, or what was now called by her husband the "inductive effect" on reversibility. Here, for the first time, kinetic effects were adumbrated in her observation that the primary effect of polar conditions was on the velocity of reversible additions, not on the equilibrium itself.

All of her remaining investigations of what would now be called nucleophilic addition effects on tautomerism were done with her husband or with shared pupils such as K. E. Cooper and Florence Shaw (who later lectured at University College in Leicester). At the same time, she engaged in a series of articles with Ingold and his pupils on electrophilic aromatic substitution and the alternating effect in carbon chains for which Christopher Ingold was to make his name through the important series of *Annual Reports* that he (and Hilda) prepared for the Chemical Society between 1924 and 1928.

By 1927 Hilda had had the first of their two daughters, Sylvia (who became a doctor) and, in 1929, their only son, Keith Usherwood Ingold, who became a distinguished chemist in his own right in Canada. In 1930, the Ingolds returned to London where Christopher headed the department of chemistry at University College. Hilda resumed chemical research in 1931 when husband and wife began work of electrophilic addition reactions. The birth of a second daughter, Dilys (later Dilys Jones) put paid to further collaboration with her husband. When however, Ingold underwent an eye operation in 1933, she took over the supervision of one of his Indian students. By then, Christopher Ingold had turned to Edward Hughes as his new principal partner in what was later to become regarded as the classic use of kinetics to interpret mechanisms. One of the exciting new tools that Hughes and Christopher Wilson took up for such research was the use of radioactive isotopes as substituents. Intriguingly, in 1934 Hilda did co-author a letter to *Nature* with her husband and their two former Leeds colleagues, H. Whitaker and R. Whytlaw-Gray, in which they showed that the proportion of deuterium oxide (D_2O) in Leeds and London tap water was one part in 9,000. The densities of the water samples were measured to the remarkable accuracy of 1 in 10^{-7}. One wonders whether Hilda did some of the analyses at home while looking after her youngest daughter. With the aid of a paid helper at home, Edith was able to travel in to University College every day to assist her husband. She would leave college early enough to be at home in Edgware for the children after they returned from school.

There are indications that she was ready to resume research in 1939 when her youngest daughter was eight since she began collaboration with the Austrian refugee chemist, Albert Wasserman, on metal sulfides as catalysts in the polymerization of olefins. This joint work was published in the *Transactions of the Faraday Society*. The Ingolds helped several Jewish refugee chemists such as Wasserman, some of whom were given temporary refuge in the Ingolds' home at Edgware. With war imminent, in 1939 Ingold's chemistry undergraduates and postgraduates, as well as a skeleton staff, were dispersed to north Wales, with general degree students being taught at Bangor and honours students at Aberystwyth. Because of their very different syllabuses, the Welsh students taking University of Wales' degrees were taught quite separately from the London students. The whole operation was masterminded from Aberystwyth by Hilda. Perhaps, not surprisingly, she also took over her husband's administrative and secretarial work with great flair. It was not until 1940 that she was granted a salary for this work, to which she became dedicated until her husband's retirement. On the department's return to a bomb-torn University College in the summer of 1944, she continued her administrative duties until 1946, when her work was given formal recognition (though because the Provost objected to husbands and wives working together in the same department the subterfuge was used of paying her salary by notionally increasing the department's annual budget by £400 per annum). There are memories of her playing tennis on the courts behind the University College laboratories. Her research career, which she might well have resumed but for the war, was at an end. But there was one exception.

In 1947 she and Ingold sent a letter to *Nature* in which they reviewed the myriad structures that had been suggested for dinitrogen tetroxide. They suggested that it was a resonance hybrid or a mesomeric molecule. The letter looked forward to the way that University College was beginning to pilot the renaissance of inorganic chemistry through its post-war ICI Research Fellow, Ronald Nyholm. At the same time, it looked back to the highlight of their husband–wife partnership—the concept of mesomerism they announced in 1933 after its adumbration in 1926.

In conclusion, as historians of women's place in society have noted, the restrictions under which early twentieth-century women chemists worked called for very considerable independence, inner resourcefulness, initiative and determination. The crystallographer, Kathleen Lonsdale (née Yardley) was able to combine motherhood and publication by using her mathematical skills at home to develop crystallographic tables for which she needed no laboratory facilities. Hilda Ingold was unable to do this. Given her Armstrongian background of high aspirations, self-learning and manual dexterity, her training at North London Collegiate and Royal Holloway, it was to be expected that Hilda Usherwood would prove an exceptionally talented woman. Her chemical career, however, tends to confirm historians' conclusions that women chemists tended to become the collaborators and assistants of male chemists unless they moved into the adjacent under-developed field of biochemistry, which Hilda did not. By changing her name to Ingold and foregoing productive research in kinetics during the 1930s while she raised her family, her extraordinarily productive husband subsumed her identity. Just as collaborating males suffer from the Matthew effect that the first-named in a collaborating set wins the praise and the awards, Hilda became Mat(Hilda) Ingold in the sense of the biblical story of Martha and Matilda.

To be fair to Christopher Ingold, however, he never failed to acknowledge his wife's help and assistance. Moreover, he always recognized Hilda's contribution to theoretical chemistry. This is marked by the fact that he continually acknowledged that their joint paper of 1926 (in which the Lowry tautomeric mechanism in alternating carbon chains became part of the *normal* structure of molecules) had been the key paper in the downfall of traditional structural theory. This was a point that he further acknowledged in the series of *Annual Reports* that appeared solely under his name.

Part 5: Chemical Books and Journals

The development of chemistry teaching and the growth of laboratories are but two indicators of how chemistry has expanded since the 1840s. Other indicators have been the advent of national chemical societies and their publications and the vast expansion of commercial science journals devoted to chemistry of which William Francis's *Chemical Gazette* was but one early example. From the late eighteenth century onwards periodicals came to largely displace chemical monographs as vehicles of new knowledge. Many of these journals, whose serializations often continued into the twentieth century, were known by their founders' names. Such eponymous journals have virtually disappeared in the twenty-first century in the interests of uniform English language texts and European integration. The first essay asks whether the disappearance of eponymous foreign language journals is a loss or a gain.

Textbooks, as well as periodicals, were a staple form of publication for various nineteenth-century publishers, as discussed in the second essay on Richard Taylor's company. Teaching manuals were, and remain, a valuable extra source of income for chemical authors. Publishers also produced reference works of chemical data or guides to the growing chemical literature. These were expensive projects and usually required collaborative effort and the financial resources of a learned society to support publication. That Henry Watts, the librarian of the Chemical Society, was able to produce a multi-volume *Dictionary of Chemistry* single-handed is a remarkable story, especially when it is realized that he also personally translated Leopold Gmelin's *Handbuch der Chemie* in eighteen volumes.

Although Robert Warington, the founder of the Chemical Society in 1841, published his account of the conditions necessary for running a successful aquarium in the *Journal of the Chemical*

Society in 1852, it was an equilibrium principle that had simultaneously occurred to others like the naturalist Philip Gosse. The latter, a member of the Plymouth Brethren, is now best remembered for his curiously titled geological treatise *Omphalos* (1857) in which he rejected the possibility of evolution on the grounds that Adam had possessed a navel even though not of woman born. The story of the formulation of artificial sea water for aquaria illustrates how chemical discussions did not always take place in the orthodox chemical literature.

The remaining four stories are concerned with books that reflect wider questions about the development of chemistry. My serendipitous discovery of a monograph on chemistry written for insurance agents in 1886 leads to reflections on how chemists and other interested parties developed ways of dealing with the risks that handling and storing chemicals may pose. Mathematics is now an essential part of a chemist's education, but it was not always so; indeed, the dominance of organic chemistry in the nineteenth century made mathematical knowledge appear superfluous. This situation was challenged by the arrival of physical chemistry at the end of that century. This rapidly led to the production of textbooks that dealt with the kinds of mathematical techniques that chemists needed. One pioneer of such texts was the pottery chemist Joseph Mellor, whose career is the subject of the penultimate story. Finally, one of the most successful series of chemical lectures has been the Baker Lectures founded in 1928 and given annually by a distinguished visiting chemist at the University of Cornell. Uniformly published in blue buckram, the elegant volumes provide a wonderful panorama of the development of twentieth-century chemistry and its preoccupations.

CHAPTER 29

The Fate of Eponymous Chemical Journals

The first periodical devoted exclusively to chemistry was Lorenz Crell's *Chemisches Journal* which was first issued in April 1778. Some eighty years later, in 1862, Carl Remigius Fresenius launched the first specialized chemistry journal, *Zeitschrift für analytischen Chemie*. The older chemical literature always cited these journals by the names of their founding editors; hence *Crell's Journal* and *Fresenius's Zeitschrift*. The same nomenclature was adopted in Great Britain with the early chemical and physical periodicals, *Nicholson's Journal* (founded 1797), *Tilloch's Philosophical Magazine* (founded 1798) and *Thomson's Annals of Philosophy* (founded 1813). Other periodicals, such as Liebig's *Annalen der Chemie*, *Berichte der chemischen Gesellschaft* (begun by the newly founded German Chemical Society in 1867), Erlenmeyer's *Zeitschrift für Chemie*, and even Kolbe's *Journal für praktische Chemie*, which accepted some of the young Wilhelm Ostwald's first publications on physical chemistry, tended to specialize in organic chemistry. In this context, the foundations of journals for physical chemistry by Ostwald in 1887 and for inorganic chemistry by Gerhard Krüss in 1892 were the inevitable reactions against Germany's love affair with organic chemistry. For this reason, no non-organic chemistry journals could ever hope to rely upon exclusively German contributions and they were forced to assume an international

The Case of the Poisonous Socks: Tales from Chemistry
By William H. Brock
© William H. Brock, 2011
Published by the Royal Society of Chemistry, www.rsc.org

character. Such an obligation suited Ostwald's entrepreneurial ambitions to make "physical chemistry" the essential foundation of general chemistry.

Ostwald's relations with his Leipzig publisher, Wilhelm Engelmann, were often tense but by the time of Ostwald's death in 1932, *Zeitschrift für physikalische Chemie* was appearing in as many as ten volumes a year. Its only German competitor was the *Berichte* [reports] of the Deutsche Bunsen-Gesellschaft für Physikalische Chemie which Ostwald, van't Hoff and Nernst had created in 1894 under the name of the Deutsche Elektrochemische Gesellschaft. (It had renamed itself in honour of Robert Bunsen's pioneering contributions to physical chemistry in 1902.) Its meetings and publications reflected its members' interest in the applications of physical chemistry to industrial processes, particularly in the area of electrotechnology, and it left theoretical developments to *Zeitschrift für physikalische Chemie*.

During the Third Reich, *Zeitschrift für physikalische Chemie* declined dramatically in size and influence as American chemists took the lead in physical chemistry, Jewish editors and contributors emigrated, or editors were politically pressurized into accepting dubious papers by "German" (= Aryan) chemists. Like all German science journals, *Zeitschrift für physikalische Chemie* ceased publication in 1944 to be revived in 1950 by the Bunsen Gesellschaft. Unfortunately, Leipzig being within the Soviet bloc meant that the periodical lost its international character. In 1954, to the infinite confusion of librarians and the future *Science Citation Index*, a journal of the same title was launched commercially from Frankfurt in what was then West Germany. By adopting English as its medium this "new series" (as it was sub-titled) regained the international character of Oswald's original journal. Despite the reunification of Germany in 1989, the two physical chemistry journals continued to be published side by side for some years with the same title.

The creation of the European Common Market in 1957 and its enlargement into the European Union led to a dramatic transformation of European chemical publications during the 1990s. The final issue of *Liebigs Annalen der Chemie* was published in December 1997. Published by Verlag Chemie (now Wiley-VCH) in association with the German Chemical Society, latterly it had been appearing in English with the subtitle *Organic and Bioorganic*

Chemistry – A European Journal. From January 1998, however, Liebig's name was abandoned and the journal transformed into the *European Journal of Organic Chemistry*. The decision was a controversial one. There was a shrinking market for German language periodicals, and despite the fact that *Liebigs Annalen* had been using English for some years, the periodical was still identified by organic chemists as a "German" journal. Publishers of journals and the chemical societies that use them found that they had to aim for an international market in which competition is intense. This, together with the new challenge and opportunities offered by European integration, led to the decision: Liebig's name was considered to be "too German" for a European chemical journal.

The change of title was part of a wider European rationalization of scientific publications. Both *Liebigs Annalen* and *Chemische Berichte* (founded by A. W. Hofmann in 1868) have become the nuclei of a European-wide merger of national chemical societies' publications, with one becoming *European Journal of Organic Chemistry* (EurJOC) and the other *European Journal of Inorganic Chemistry* (EurJIC).

Germany, France, Italy, Belgium and the Netherlands have sacrificed their national chemical journals, and by 1998, some 14 European chemical societies had agreed to merge their national language journals into English language publications. Editorial guidance for these publications is currently managed by a committee unimaginatively called ChemPubSoc Europe. Under its auspices the revived West German version of Ostwald's *Zeitschrift für physikalische Chemie* became part of the *European Journal of Chemical Physics and Physical Chemistry* published by the international commercial firm of Wiley. At the same time, Ostwald's original Leipzig *Zeitschrift* was bought by the Bremen publishing house of Oldenburg-Verlag and continued as a highly successful physical chemistry journal. Further rationalization took place when the *Berichte* of the Bunsen-Gesellschaft was acquired by the Royal Society of Chemistry and absorbed into a new European journal entitled *Physical Chemistry and Chemical Physics* (PCCP).

Historians of chemistry watched these changes with mixed feelings. While saddened by the loss of a great name, we might reflect that in all likelihood Liebig would have approved, given his abortive attempt in 1837 to internationalize *Annalen der Chemie* by co-opting as editors Thomas Graham in Britain and Jean-Baptiste

Dumas in France. After all, it was only after his death in 1873 that *Annalen* was renamed *Liebigs Annalen*. Historiographically, the changes prompt reflections on past national rivalries and the new spirit of internationalism abroad in twenty-first century science as English becomes the *lingua franca*. Historians will have to come to terms with the fact that the days of eponymous science journals, once essential to identification in the first half of the nineteenth century, do seem numbered. The Royal Society of Chemistry still sectionalizes the *Journal of the Royal Chemical Society* into the *Dalton, Faraday* and *Perkin Transactions*—long may that prevail.

However, there are commercial journals that have bucked this trend. *Fresenius' Zeitschrift für analytische Chemie* began to accept papers in English in the 1950s. From 1990 onwards all contributions were published in English and the title anglicized to *Fresenius Journal of Analytical Chemistry*, thus perpetuating its founder's name. In the field of biochemistry *Hoppe-Seyler's Zeitschrift für physiologische Chemie* was changed to *Biological Chemistry Hoppe-Seyler* in 1985, although the significance of Hoppe-Seyler's name is probably lost on the present generation of biochemical researchers. Russian chemists have also demonstrated that an eponym can be made to survive for proud nationalistic reasons. In 1991 the Russian Academy of Sciences, in association with the Royal Society of Chemistry, launched *Mendeleev Communications* with English as the sole language of publication for its mainly Russian contributors.

Eponyms also continue to exist in two journals devoted to the history of science and medicine, namely *Sudhoffs Archiv, Zeitschrift für Wissenschaften* and the *Transactions of the Newcomen Society*. It seems unlikely that their titles will change despite globalization.

CHAPTER 30

The Lamp of Learning: Richard Taylor and the Textbook

We would miss a great deal if we simply defined textbooks in terms of books that helped students to pass examinations. Nor is it much help to say a textbook contains the normal science of a Kuhnian paradigm since we find that nineteenth-century popular books written for women and children also discuss scientific ideas as if they are fixed and finalized. More useful, perhaps, is the model of a "thought collective" and overlapping circles of influence between specialists and laymen as defined by Ludwig Fleck in his 1935 volume *Genesis and Development of a Scientific Fact*. Historians of science would probably now prefer to translate the original German title as *Genesis and Construction of a Scientific Fact*. Fleck's concentric circles generate four types of scientific literature: an esoteric circle of primary periodical literature for the cognoscenti; handbooks and data books for general experts; textbooks for tyros who are being initiated into the esoteric circle; and popularizations for the exoteric layman. Fleck's point was that scientific information permeates both from the esoteric to exoteric circles and back again, and while doing so, information is socially and culturally consolidated and constructed into scientific facts or a thought collective. So, for whatever purpose or audience a scientific book is designed for or aimed at, its author, publisher and printer are essentially engaged

The Case of the Poisonous Socks: Tales from Chemistry
By William H. Brock
© William H. Brock, 2011
Published by the Royal Society of Chemistry, www.rsc.org

in the genesis and construction of scientific information. Publishers and printers are presumably engaged in this activity for mainly commercial reasons, albeit, those like the nineteenth-century printer Richard Taylor (1781–1858) may have had genuinely altruistic reasons, namely that they believed that science brought people nearer to God, or that a scientifically informed and educated public was the shape of things to come. As a Unitarian, Richard Taylor would have sympathized with both of these viewpoints (Figure 30.1).

The family publishing house of Taylor & Francis became a public company in its bicentennial year (1998) and celebrated the event by buying up Routledge (a spring chicken of 150 years) for £90 million. Today, the firm has a combined portfolio of over 1,000 journals (by no means all scientific ones) and publishes some 1,800 books a year and holds a backlist of some 20,000 items. It no longer prints its own material, so it is hard to imagine that this international publishing enterprise began as a small family business in a London alleyway with a single printing press and only one journal, the *Philosophical Magazine* that Alexander Tilloch began in June 1798.

Within fifty years of its foundation in 1783, Taylor's firm (Davis, Taylor & Wilks in 1800, Wilks & Taylor in 1801, and R. Taylor by 1802) had become London's (indeed Britain's) largest scientific printer and publisher. By 1833, housed in a picturesque building in Red Lion Court off Fleet Street, its printing portfolio included *Philosophical Magazine*, the annual bulky *Reports of the British Association for the Advancement of Science*, and all publications of

Figure 30.1   Richard Taylor (1781–1858), the Unitarian printer and publisher who founded the firm of Taylor & Francis and specialized in the publication of scientific periodicals. (Wellcome Images)

the University of London, the Transactions of the Royal Society, Geological Society, Zoological Society, Horticultural Society, Royal Botanic Society, the Linnaean Society (right up to 1950), the botanical and zoological catalogues of the British Museum, as well as what are today considered significant books, such as the publication of Faraday's collected papers on electricity and magnetism, and the bird books of John Gould. By the 1840s, Taylor was both proprietor and chief editor of *Philosophical Magazine* and its biological counterpart *Annals of Natural History*. Arguably, then, the firm was playing and continued to play a vital role in the dissemination of Victorian science, including impressive support for natural history.

Taylor, who was born in 1781, was a member of the talented Taylor family of Norwich, a dynasty that formed part of the intellectual and radically nonconformist aristocracy of eighteenth- and nineteenth-century England. Richard Taylor's great-great grandfather was a timber merchant in Lancashire, but his son (Taylor's great-grandfather) was the famous nonconformist divine, John Taylor, who taught at the Warrington Academy before moving to Norwich in 1733 to preach at the Octagon Presbyterian chapel. His son (Taylor's grandfather), another Richard Taylor, became a wool merchant in Norwich and forged a marriage alliance through another Unitarian family, the Martineaus. In the second half of the eighteenth century, the Taylors and Martineaus became Norwich's leading families in business, religious and civic affairs, and these duties were continued by Taylor's father, John Taylor, a wealthy wool factor, political radical and hymn writer. His wife Susannah, Richard Taylor's mother, was extremely talented and intelligent and held a literary and artistic salon in the Taylor household. Not surprisingly, John and Susannah Taylor produced seven talented children, all united by their Unitarian faith in a belief that there was a human obligation to benevolent and useful exertion during their earthly existence.

Richard Taylor, the second born of these talented children, was educated at a local dissenting school, where he seems to have acquired his interest in languages, particularly Anglo Saxon. There was clearly hope that he would aspire to the dissenting ministry but in the event, like the four other sons, Richard was apprenticed at the age of fifteen, in his case to the printer Jonas Davis in Chancery Lane, London, on 7 March 1797. The intermediary who arranged this for the parents was none other than the botanist James Edward Smith who had founded the Linnaean Society nearly ten years before. The Davis–Smith connection was, of course, that Davis was the printer for the Linnaean Society's *Transactions*, first issued in 1791. It was through this happy accident, coupled with the fact that Davis began the printing of Alexander Tilloch's monthly *Philosophical Magazine* in 1798 that Richard Taylor was to become a scientific printer and editor.

Although only in his forties, Jonas Davis decided to abandon printing in 1800 by handing over the business to his talented apprentice. But since the apprenticeship would not expire until 1804 and Taylor was not twenty-one until 1802, the firm was bought by Taylor's father who acted as sleeping partner while another young printer, Richard Wilks, was brought in to partner Richard Taylor. Unfortunately, the young men did not get on and literally came to blows in 1803, with the result that Richard Taylor moved to new premises in Black Horse Court. The partnership was dissolved and fortunately customers' good will retained by the Taylors—including the *Philosophical Magazine* and the Linnaean Society's *Transactions*. A year later, at his majority, Taylor became his own master printer and publisher, albeit continuing to rely on some fatherly financial support—notably for the development of the steam press with Thomas Bentley, George Woodfell and Frederich Koenig between 1807 and 1818.

This significant partnership demonstrates how Taylor was at the forefront of technical innovation in printing. The mechanized steam press, together with stereotyping and the repeal of taxes on stamp, paper and advertising during Taylor's lifetime was to make possible the production and dissemination of cheaper, but still profitable, well-illustrated textbooks. Taylor's colophon, the Roman lamp of learning, "Alere flamman", was first used in 1813 and aptly symbolizes his commitment to the scientific and intellectual community. He first moved his presses into the famous Red Lion Court premises off Fleet Street in 1827.

Taylor's private life in this early period was equally dramatic. In 1807 he married a London tailor's daughter, Harriet Corke, by whom he had a daughter Sarah. A seventeen-year old governess and printer's daughter named Frances Francis was employed in 1812 and the inevitable happened. Whether his wife Hannah was already mentally unstable, or became mentally disturbed as a result of this adulterous liaison, we do not know. Suffice to say that to all intents and purposes Frances became Taylor's wife and rapidly bore him two children, William and Rachel Francis. In the eyes of Taylor's family, Taylor's behaviour was outrageous and had to be hushed up. The two Francis children were brought up secretly in Hastings until 1833 when they moved to Clerkenwell close to Taylor's home in Charterhouse Square to which he had moved after his wife's death; it would appear that she spent her last years in a private lunatic asylum. Taylor ensured that William Francis had an excellent education on the continent at St Omer's College, with further studies at the universities of Berlin and Giessen, where he obtained a doctorate with Liebig in 1841. During his studentship, Francis hit upon the idea of a chemical periodical that would provide rapid translations and abstracts of French, German and Scandinavian research. His father agreed and the result was the *Chemical Gazette* which appeared fortnightly between 1842 and 1859. Taylor took Francis into partnership in 1851. This partnership led to terrible row with his two engineering brothers, John and Edward Taylor, who succeeded in having Richard certified in May 1852. It was probably a nervous breakdown that justified this drastic action, which did not however affect the legality of the partnership with his natural son. A year later, Taylor was rescued by his daughter Sarah and William Francis (who were on the best of terms) and moved to a house in Richmond, where he died 1 December 1858.

Until the 1840s at least, Richard Taylor was active in Unitarian affairs, worshipping at the chapel in Southwark. He published and printed tracts arguing for the scriptural authority of Unitarianism and he printed at least one petition in favour of the repeal of Acts of Parliament that opposed Unitarianism. He also supported Catholic emancipation. A great admirer of the chemist and theologian Joseph Priestley, he reprinted Priestley's *Lectures on History* (1788) from an American edition of 1826 and supplied notes to the edition with his friend J. T. Rutt.

Taylor took an active part in the life of the City of London where, for 35 years, he was a Common Councillor for the ward of Farringdon Without. He was particularly active in the cause of education. He was a member of the Friends of Universal Education and was involved in the foundation of both the Harp Alley Boys' School and the City of London School in 1834. He was one of the promoters of the new University of London in 1828, and in 1834 he was the chief member of the Common Council to petition William IV for a Royal Charter of Incorporation for the university so that it could award degrees. The Privy Council negotiated directly with Taylor in changing the University Deed of Settlement whereby none of the original proprietors of the University (like Taylor himself) could benefit financially from their shares. When the King granted the university its charter in November 1836, it was Taylor who printed it. Two years later (1838), Taylor was appointed official printer to the University of London (on condition that he charged no more than the official government printers). In justifying his appointment he wrote of his "experienced proof correctors" and the extensive stock of "Hebrew, Greek, Oriental, Gothic, Saxon, German [Fraktur] and other types" and his "considerable library for the facility of reference, both philological and scientific". Most of his official university work was evidently concerned with printing regulations and examination papers. There exists an undated letter from a student offering Taylor money for a copy of questions on the forthcoming matriculation exam, adding that "consequences of unusual interest depend on whether I pass or not".

It must be emphasized that Taylor did not publish textbooks for the university. Confusingly, this was the job awarded to John Taylor (1781–1864) of Taylor & Hessey in Fleet Street who became the publisher (as opposed to printer) to the University of London in 1836. This firm, best known for publishing the work of the poet John Keats, had moved to Upper Gower Street in 1841 and become Taylor & Walton (later Walton & Maberly). It was this firm that began to specialize in textbook production for schools and universities. They were also Liebig's English publishers. Confusingly, Richard Taylor frequently printed these texts.

Taylor's records in the St Bride Printing Library for the period 1813–1832 log the printing of dozens of texts for University College London (*e.g.* Spanish, Latin and Hebrew Grammars). In a typical week in 1831 we find Taylor printing Samuel Parkes, *Chemical Catechism*, school prize exercises for Charterhouse, "Latin Subjunctive" and J. F. Johnston's best-selling *Chemistry of Common Life*, as well as masses of examination papers for middle class schools such as Charterhouse, Cheltenham, Wimbledon, Stoneyhurst, King Edwards Birmingham and University College.

In general, the texts always tended to be rather specialized ones calling for unusual typefaces; for example, Frederick Accum's *Chemical Amusement* (1818), J. Bosworth's *Anglo-Saxon* (1822), C. J. Blomfield's *Greek Grammar* (1827) and, for the publisher John Van Voorst, Edward Frankland's *Lecture Notes for Chemical Students* (1866), which was the first textbook to use graphic chemical formulae. Taylor did publish simpler texts, such as the anonymous text that Elizabeth Fry compiled in 1830 for her prison school. However, by the 1870s, the firm was printing fewer books because it had become a specialist periodicals printer and publisher.

Taylor's most important printing job was undoubtedly the work of the Unitarian industrial chemist, Samuel Parkes (1761–1825). Parkes distributed his books through various publishers, but he used Taylor as his printer from 1805 until his death, leaving Taylor as his executor and copyright owner. Consequently, Taylor continued publishing new editions of *The Chemical Catechism* (first published in 1806) up to 1834, and third and fourth editions of Parkes's *Chemical Essays* (first published in 1815) up until 1841. Taylor only ceased printing new editions of these works when it was no longer worthwhile updating Parkes. There were more systematic textbooks available once chemistry had become a teaching and examinable subject in schools and universities.

A sure sign of the success of Parkes's formula was the rapid appearance of pirated editions of *The Chemical Catechism*. One of these, the *Grammar of Chemistry* by Sir Richard Phillips, appeared for the school market in 1809 and led Parkes to reconstruct the *Catechism* as a school text, *The Rudiments of Chemistry*, in 1810. The four editions of this didactic, non-catechetical text were again printed by Taylor. It would prove a meaningful exercise in textbook design to compare the original *Catechism* with the *Rudiments*.

Taylor was never a systematic publisher of textbooks, but an opportunist who believed in education, enlightenment and science and who, in his own skilled way as a scholar printer, contributed to what has been called the construction of scientific facts.

CHAPTER 31

"The Greatest Work which England has ever Produced": Henry Watts and the *Dictionary of Chemistry*

What have been more useful to practising chemists: textbooks, monographs or reference works? The chemist William Crookes had no doubt that the last were the most valuable. When he reviewed the complete five-volume set of Henry Watts's *Dictionary of Chemistry* in 1868 he described it as "the greatest work which England has ever produced in chemistry, [indeed] one of the greatest she has produced on any subject". All the more remarkable, then, that the first edition was more or less the work of one man (Figure 31.1).

The "greatest work" was but one of Henry Watts's publications for chemists. Watts was born in London in 1815 and apprenticed to a surveyor when he was fifteen years old. Dissatisfied with his career, he began a long period of self-education while working as a private tutor. After graduating from the University of London in 1841, he assisted George Fownes, the Professor of Analytical and Practical Chemistry at University College's Birkbeck Laboratory, where Thomas Graham, a founding member and the first president of the Chemical Society of London, was the principal professor. In 1847 Watts revised Graham's *Elements of Inorganic Chemistry*, and after Fownes's tragic early death in 1849, he also revised several posthumous editions of Fownes's *Manual of Chemistry*.

The Case of the Poisonous Socks: Tales from Chemistry
By William H. Brock
© William H. Brock, 2011
Published by the Royal Society of Chemistry, www.rsc.org

Figure 31.1 Henry Watts (1815–1884), an assistant to Alexander Williamson at University College London who became librarian at the Chemical Society in 1849. Most of his life was spent in valuable editorial projects. (RSC Library)

Meanwhile, Watts continued to assist Fownes's successor, Alexander Williamson. But for a speech impediment, Watts would undoubtedly have eventually found an academic position. But since he did not wish to spend his life as a laboratory assistant, he turned instead to writing and librarianship. In 1848, at the request of the Cavendish Society formed by Thomas Graham as a subscription club to publish works of value to chemists, Watts began to translate and to edit the most recent edition of Leopold Gmelin's massive *Handbuch der Chemie*. The translation of the eighteen volumes was not completed until 1872 and remains today of the greatest value for historians of nineteenth-century organic chemistry. Backed by Graham and Williamson, Watts became editor of the London Chemical Society's *Journal* in 1850 and its librarian in 1861. He held both positions until his death in 1884.

In 1858, when Watts was in the midst of the daunting task of translating and updating Gmelin's text, the publisher William Longman commissioned him to revise Andrew Ure's one-volume *Dictionary of Chemistry and Mineralogy* (1821). Ure's *Dictionary*, itself a rewrite of William Nicholson's *Dictionary of Chemistry* of 1795, was so seriously out of date that Watts decided to begin again from scratch. It is said that only nine pages of the eventual five volumes contained the original text of Ure's fourth edition of 1835, though Ure's influence is evident in the amount of space given to mineralogy. The new *Dictionary of Chemistry and the Allied Branches of Other Sciences* was issued in monthly installments of 192 pages (twelve sheets of sixteen pages), beginning in May 1863. William Francis, editor of the monthly *Philosophical Magazine* and a pupil of Justus von Liebig, thought that the parts were too large to guarantee adequate proofreading. He urged Watts and his publisher to reduce the monthly fascicules to six sheets even if it lengthened the publication process. Francis's advice went unheeded. The first volume was over a thousand pages long and the errata list was, indeed, a lengthy one.

Watts commissioned a group of underemployed London chemists to help him with the project (though at least three quarters of the *Dictionary* was his alone). Several of the helpers, like Edmond Atkinson, George Carey Foster and Frederick Guthrie, all of whom had studied chemistry at German universities, eventually found careers in physics rather than in chemistry. William Stanley Jevons (who wrote entries on the balance, the barometer, and clouds) abandoned chemistry completely and became famous as an economist. Hospital chemists such as William James Russell and William Ditmar contributed, as did William Odling (then employed at St Bartholomew's Hospital) who completed a significant entry on atomic weights in which he extolled the system of the French chemist, Charles Frédéric Gerhardt. Ongoing changes in chemical knowledge left their mark on the work. The atomic weights of many elements doubled while the translation was in progress, and to distinguish these amended values from the earlier ones Watts adopted double symbols such as Bba = 137 (barium) and Hhg = 200 (mercury), only to abandon them in the fifth volume as too clumsy.

The *Dictionary* went rapidly out of print. By its very nature, it needed updating immediately after the first edition was published; internal evidence suggests that the articles covered the literature up to 1861. Confusingly, a corrected "new edition" appeared between 1874 and 1875, even as a genuine second edition was appearing in five volumes over the years 1872 to 1877, together with three unwieldy *Supplements* issued in 1872, 1875 and (in two volumes) 1879–1881. By 1881 Watts's various contributors had become distinguished men and the title page phrase "assisted by eminent contributors" was undoubtedly true.

Although he did not produce original research, Watts was elected a Fellow of the Royal Society in 1866 on the strength of his many contributions as an editor and translator. (He also found the time to translate from the French Charles Wurtz's *History of Chemical Theory from the Age of Lavoisier* in 1869.) Friends also successfully lobbied the government to award Watts a civil list pension of £100 a year in 1876. He died in June 1884 in the middle of revising a translation of Friedrich Knapp's *Technologie* that Knapp and Thomas Richardson had first prepared in 1847. The work was eventually published in 1889 as *Chemical Technology* under the editorship of William Thorp and Charles E. Groves, Watts's successor as editor of the *Journal of the Chemical Society*.

The *Dictionary of Chemistry* proved too useful to abandon with Watts's death and Henry Forster Morley and Mathew M. Patterson Muir were soon commissioned to revise it; the revision appeared in four volumes over the years 1888 to 1894. Meanwhile, inspired by Watts's successful formula, Thomas Edward Thorpe, who was Government Chemist and had been a pupil of Bunsen's in Germany, compiled a parallel three volume *Dictionary of Applied Chemistry* in 1893. This rapidly became a major source of reference, and its succeeding three twentieth-century editions replaced Watts's *Dictionary* completely. A fourth edition of *Thorpe's Dictionary* was greatly affected by the Second World War, production taking from 1937 until 1956. The delay may well have inspired the American chemical engineers Raymond Eller Kirk and Donald Othmer to compile their *Encyclopaedia of Chemical Technology* in seventeen volumes between 1947 and 1960.

The lineal descendant of Nicholson's and Ure's modest chemical dictionaries and Watts's staggering bibliographical contributions to Victorian chemical information, *Kirk–Othmer* remains in print. Crookes was right in underlining the usefulness of such encyclopaedic contributions to chemical literature. What he overlooked, however, is that as they grow out of date they become of great value to the historian.

CHAPTER 32

Chemistry in the Aquarium

The aquarium craze which swept through Great Britain in the 1850s followed on from what has been described as a fern craze in the previous decade. In 1829, the London physician and botanist Nathaniel Bagshaw Ward (1791–1868) invented the "Wardian case" for transporting growing plants from one geographical place to another. The case simply exploited an equilibrium principle that had already been noticed by Joseph Priestley when he had investigated photosynthesis at the end of the eighteenth century. In the sunshine, transpired water evaporates and during the night condenses on the leaves, while the input and output of carbon dioxide and oxygen remains in equilibrium by day and night. The aquarium was a logical development of Ward's device which had been taken up avidly by fern collectors; take a glass tank half-filled with water and establish an equilibrium between aquatic plants which give off oxygen in light to aerate the water, add a few fish and marine creatures, together with scavengers such as snails and anemones to remove decomposing matter, and, in principle, the water never need be changed. To most religious Victorians the aquarium was a welcome illustration of the divine order and harmony of creation (Figure 32.1).

In one of the nine papers on the development and management of the fresh- and sea-water aquarium, Robert Warington (1807–1867), the founder of the (Royal) Society of Chemistry in 1841 and

The Case of the Poisonous Socks: Tales from Chemistry
By William H. Brock
© William H. Brock, 2011
Published by the Royal Society of Chemistry, www.rsc.org

Figure 32.1 Robert Warington (1807–1867), an assistant to Edward Turner at University College London, he worked as a brewery chemist before becoming the chief chemical operator at the Society of Apothecaries in 1842. He was the effective founder of the Chemical Society in 1841 and developed the aquarium in 1850. (RSC Library)

the Chemical Operator of the Society of Apothecaries in London, confessed that although he had been successful in obtaining fresh sea water from oyster stall-holders at Billingsgate Fish Market, "for a considerable period after commencing these [sea-water] experiments [in 1852], I was much troubled to obtain living subjects in a healthy condition". His difficulty had been solved through the kindness of the naturalist Philip Henry Gosse (1810–1888), whose independent paper on the principle of a sea-water aquarium had appeared simultaneously with Warington's in the October 1852 issue of *Annals and Magazine of Natural History*. Coincidentally, both authors had dated their contributions 10 September 1852. Gosse, who had immediately conceded priority of publication to Warington had, Warington publicly acknowledged, "offered in the kindest manner possible to supply me with materials, and from that period he has always most heartily responded to my wants". Indeed, Gosse had offered to supply Warington with red algae of various species and *Medusae* in a growing state and send them to London from Devon by rail if Warington would only supply jars with sealable lids.

Such concord was not to last. Behind the scenes the Weymouth solicitor and naturalist, William Thompson (1822–1879), who was a regular contributor to *Annals and Magazine of Natural History* and (according to Gosse) an important supplier of specimens, informed Warington that in his view Gosse's work was completely unoriginal and that Warington was undoubtedly the real creator of "the vivarium".

"I am sending you a copy of my communication to the *Annals of Natural History* for this month. I need not apologise for sending this as I consider you the first projector and originator of the Marine Vivarium [*i.e.* aquarium]. After reading the *Chemistry of Creation* by Robert Ellis 1850 in reference to the Chemistry of the Oceans and your papers in the *Gardeners Chronicle, Chambers Journal* and the *Annals*, I hesitate very much in the propriety of according to Mr Gosse any honour for the originality of his paper in the *Annals* for October 1852. You will see that I had found the judicious balance of animals and vegetable life protected the life of both, but being no Chymist I did not trace it to its right source and therefore would not assert it as a fact not yet received. I have great facilities for carrying out experiments in this matter but am debarred from doing so by my ignorance of Chymistry of the Air and Ocean beyond the main threshold. Should you, however, feel disposed, I think your knowledge of Chymistry combined with my knowledge of Nat. Hist. and facilities for carrying it out, might, if we work together produce important results."

In his enclosed *Annals* paper, Thompson explained how he had stumbled upon the marine aquarium principle in 1850 after first changing the sea-water in which his littoral species were preserved twice daily "to represent the tides". Struck by the beauty of the fronds of storm-tossed seaweeds attached to pebbles on the Weymouth shores, he had serendipitously placed them in his sea-water vessels containing *Crustacea, Echinodermata, Testacea* and *Zoophytes*.

"I watched them daily to change the water as soon as I detected my prisoners becoming sickly and (with the exception of one or two that died and which I removed) to my astonishment at the end of a month the whole of the animals were healthy, and the water remained in my opinion pure and limpid. This fact, through my ignorance of 'Chemistry of Creation', I did not set down to the right cause."

Clearly, Thompson posed no priority threat to Warington. In any case, the main purpose of his paper was to proffer advice on the construction of large wooden "vivariums" for use "as ornaments to the drawing room or for the purposes of instruction in museums, Literary Institutions [presumably literary and philosophical societies], and perhaps even the school-room".

There is no evidence that Warington ever collaborated with Thompson. Like Gosse, he must have acted as a further stimulus to naturalists who had already been inspired to build aquaria after reading abstracts of Warington's original *Chemical Society* paper in the journals mentioned by Thompson, as well as others like *The Zoologist* and *The Florist*. Given the promulgation of physico-chemical circulation theology in the 1830s and 1840s, which provided a basis for agricultural and sanitary reforms as well as the subject of Ellis's book, together with the introduction of the Wardian case for transplanting plants and the repeal of the tax on glass in 1845, the simultaneous development of the marine vivarium, or aquarium, by several workers is scarcely surprising. However, despite interesting empirical developments during the 1840s by Nathaniel Ward and Mrs Anna Thynne, Warington was undoubtedly the first to develop the aquarium from theoretical principles. Nevertheless the matter was then controversial and Warington was clearly anxious to reserve credit for himself.

Whether because of Thompson's private intervention or of jealousy at the way in which Gosse, as a professional naturalist, succeeded in creating the commercial wave of "aquarium mania" with popular texts and the commission to stock the London Zoological Society's aquarium, Warington did not remain on good terms with Gosse.

In July 1854, back in London, Gosse described a simple way of preparing artificial sea-water, using the analysis of sea-water published by the Brighton mineral water manufacturer, Schweitzer.

In December, Warington (a professional analyst) described Gosse's formula as "erroneous". Gosse's ignorance of chemical analysis, he stated, had led him to make three errors. He had miscalculated the proportions of the salts necessary, he had assumed incorrectly that Schweitzer had referred to crystalline magnesium sulfate (Epsom salts) when, like all analysts, he had referred to the anhydrous form, and he had used more water than was proper by failing to allow for saline volume in his standard volume. For good measure he also claimed that Gosse had consulted him about the feasibility of making artificial sea-water in January 1854.

> "I told him that there could be no difficulty in the matter, as I had made and had then in use several small quantities artificially produced, and that all was required was that a good analysis should be taken as the basis for deducing the proportions, and at the same time referred him to the source which I myself had worked, namely Dr E. Schweitzer's analysis of the water of the English channel taken off Brighton."

A note in Warington's handwriting in the archives of the Rothamsted agricultural research station lists all the visitors who came to see his aquarium. This confirms that Gosse visited Warington on 16 and 21 January 1854. Nevertheless, in *The Aquarium*, published later that year, Gosse continued to advise London aquarists to pay "a trifling fee to the master or steward of any of the steamers that ply beyond the mouth of the Thames, charging him to dip [a 20 gallon cask] in the clear open sea beyond the reach of rivers".

In fact the differences between Warington's and Gosse's formulae for artificial sea-water were not all that significant. Warington used slightly greater amounts of sodium, potassium and magnesium chlorides and magnesium sulfate, and also included the magnesium bromide and calcium carbonate and sulfate which Gosse had considered too insignificant to include. Because the major factors in the preparation of sea-water for aquaria are the total salinity and hence specific gravity, it is scarcely surprising that Gosse's much simpler recipe worked. He was writing for "practical people, to whom minute accuracy was impossible", Gosse informed readers of the *Annals* in reply to Warington's challenge. The fact that Gosse's formula was scientifically inaccurate did not convict him of being a poor scientist. On the contrary, he declared, he had proved the safety of his formulation before publishing it, for the sea-water he had prepared on 21 April 1854 and written about in June was still supporting life. As to Warington's claim that he had merely followed up advice given him by Warington, he retorted angrily: "If Mr Warington supposes that I obtained from him one atom of information unknown to me, on the subject of making sea-water from its constituent salts, he is most thoroughly mistaken. He is no less wrong in saying that I 'consulted' him, since I merely mentioned what was in my mind in familiar conversation". And there the matter ended.

Why did Warington and Gosse fall out? In many ways they were ideal collaborators and Gosse certainly fulfilled Warington's plea for trained naturalists to exploit the aquarium as a way of studying the mysterious denizens of the aquatic world. Gosse had direct access to specimens, a wonderful flair for scientific description and biological expertise; Warington had chemical expertise and was the intimate friend of almost every physical scientist in London. Both men shared a religious conviction that the aquarium was a microcosm, or parable, of a divinely ordained and harmonious cyclic macrocosm. What, then, went wrong in the relationship between the solitary Gosse and the gregarious Warington?

In the absence of a written statement from either man or their contemporaries, we can only speculate. But the answer is probably quite simple: Warington was jealous of Gosse's success in a field in which he had made the major scientific breakthrough. Warington, a busy metropolitan practical chemist with heavy day-to-day responsibilities for the standardization and prevention of adulteration of drugs, could only turn to his aquarium research as a leisure activity. It was Gosse, whose livelihood depended upon literary activities, who had the time and literary ability to write the best-selling *A Naturalist's Rambles on the Devonshire Coast* (1853) and *The Aquarium* (1854); it was Gosse, not Warington, whose work formed the focus of Charles Kingsley's *Glaucus* in 1855.

Although both Gosse and Kingsley gave Warington prior credit for the balanced aquarium concept in 1850, Warington must have felt nevertheless that, unfairly, through its commercialization and popularization, the aquarium had become more associated with Gosse's name than his own. As Warington's surviving jottings demonstrate, he became obsessed with the demonstration of his own priority. Gosse's claim in July 1854 to have "invented" an artificial sea-water at a time when the newly published *The Aquarium* was receiving adulatory reviews only seemed to confirm what Thompson had suggested the year before. The result was Warington's bitter note on Gosse's sea-water paper in the December 1854 issue of *Annals of Natural History*.

CHAPTER 33

Insurance Chemistry

Browsing through any large collection of books always produces at least one title that intrigues and gives pause. We all know that chemistry has diversified into myriads of tiny specialisms. But who among us who does not work in an insurance office would expect to find a book entitled *Harris's Technological Dictionary of Insurance Chemistry*?

This interesting compendium was first published in Liverpool in 1886 by the Phoenix Fire Insurance Company under the title *A Technological Dictionary of Fire Insurance*. Its 407 pages had alternate pages left blank for the owner to write down notes and additional information. The author, William A. Harris, was the secretary of the Phoenix Insurance Company, one of the oldest insurers in Great Britain. It had been founded in a London Coffee House in 1781 to insure sugar refiners against losses caused by fires and gradually expanded into other areas of insurances against risks. Initially, business outside London was conducted by means of locally appointed agents, but in its centenary year, provincial offices were established in Glasgow and Liverpool and other major ports. Harris, a former army colonel who had been in charge of local agents at the firm's London headquarters in Lombard Street was appointed secretary of the Liverpool branch when it opened for business in 1885. The official historian of the present day Phoenix Assurance Company described Harris as "an organizer, a Fellow of several learned societies in the United Kingdom and

The Case of the Poisonous Socks: Tales from Chemistry
By William H. Brock
© William H. Brock, 2011
Published by the Royal Society of Chemistry, www.rsc.org

in the United States, [and] he was a sound fire insurance underwriter".

The *Technological Dictionary of Fire Insurance* was addressed to insurance managers, underwriters and insurance agents; surveyors and government officials; captains, shippers, stevedores and harbour masters; railway, dock and traffic managers; merchants and warehouse keepers—to explain the various risks from "spontaneous combustion, oxidation, chemical affinity, fermentation, friction, expansion of gases, inflammability of vapours, dust explosions, steam heating and drying, oils, fibres, coal, cotton, and mixed cargoes, etc."

There is internal evidence that Harris had lived in Philadelphia for a time since there are quite a few references to American chemical fires and explosions, and to his membership of Philadelphia's Franklin Institution and to Boston's Society of Arts (the future Massachusetts Institute of Technology). It seems, for example, that Philadelphian gentlemen in the 1880s lit their cigars with pieces of sodium! Clearly the book is the equivalent of today's list of hazardous chemicals.

By 1890, Harris's *Technological Dictionary* was out of print and this prompted him to recast it completely as *Harris's Technological Dictionary of Insurance Chemistry*, the revised title emphasizing the chemical nature of the hazards and risks that were described. Harris now saw his readership as worldwide. The book was a "compendium of the latest information on subjects of vital importance to Insurance Managers, surveyors, underwriters, government officials in all countries, merchants, captains, shippers, harbour-masters, railway, dock and traffic managers, warehouse-keepers, stevedores, manufactures, home, colonial, and foreign insurance agents, explaining the various risks to be apprehended" in handling chemicals and natural products. In a verbal extravaganza, the book drew attention to the careless ways of:

"loading, stowing, and conveying by sea; and of manufacturing, packing, baling, casing, caring, carrying, sheltering, warehousing, and general handling and treatment on land, of goods, and merchandise ... which, though possibly in themselves of a non-hazardous nature, may—from amalgamation, massing, chemical combination, compression, contiguity, friction and treatment, or accidental impregnation, with pigments, dressings, acids, emulsions, oils, gases, or vapours; dyeing, drying, improper lubrication, breakage, leakage, or other causes—set up one of other of the dangerous conditions [to be] enumerated."

Four years later, in 1894, Harris produced his final manual, *A Technological Fire Insurance Commentary*, which he claimed superseded his previous two volumes. Entries ran from "abattoirs", which were a risk because of the wooden louvre boarding that was commonly used in construction (as well as from the class of men employed) to "zinc water", a cheap way of using zinc hydride as a fireproofing agent. Zinc metal itself was never to be used to encase a fire door because it had a low melting point. Harris listed the hazards associated with a huge range of chemicals and raw materials such as alpaca wool whose dust could explode, as well as particular building types such as arcades (where cigarette and cigar ends were a fire hazard in confined quarters) and wooden bookstalls at railway stations where waste paper proliferated. We learn that coal fires were common in London docks, of the hazards of cotton fires on ships, and the real dangers of milling paper. He did not hesitate to mention the apparently trivial danger—the Sun's rays on chemists' bottles. These were the days when chemists' shops attracted custom by displaying large bottles of beautifully coloured liquids such as copper sulfate. One chemist had failed to draw the shop blinds on a particularly sunny day. In his window was a large carboy of potassium dichromate which exploded. Not surprisingly, Harris also devoted many pages to fire-extinguishing solutions such as one made up from four parts of iron sulfate and sixteen parts of ammonium sulfate in 100 parts of water. This was supposed to work by producing large volumes of vapour (ammonia) that excluded the air necessary for combustion.

Governments and shipping insurers were put on their toes between November 1865 and April 1866 when there were a series of explosions of Alfred Nobel's new blasting agent in New York, San Francisco, Panama and Sydney. Inquiries revealed that consignments of Nobel's nitroglycerine from his factory at Kummel in Germany were frequently misleadingly labelled and that no special precautionary instructions had been given for the material's handling and storage. This was no doubt a deliberate attempt to evade higher freight charges for dangerous goods that required special and careful handling. State legislators, shipping owners soon banned the movement of the new explosive; for example, in August 1869 the British government banned the importation of pure glycerine.

Such prohibitions might have proved fatal for Nobel's business, but he had been made well aware in May 1866 when his nitroglycerine factory exploded, that he had to find a way of taming the dangerous liquid. He rapidly solved the problem by absorbing it in kieselguhr, an absorbent fine-grained rock. In this form, which he named "dynamite", the explosive was safe unless detonated by a weaker explosive. Nobel rebuilt his factory and made his fortune with the safer product.

By the end of the nineteenth century, insurance had become completely professionalized. Insurance societies and institutes proliferated in provincial towns and cities. In 1897 it was suggested that they should form a federation and examination body, as well as publishing a journal. By 1912 the Federation had become the Chartered Insurance Institute, with fire insurance forming one of the three examination subjects alongside life and accident insurance.

That Harris's manuals filled a need is demonstrated by the lectures provided towards the end of the nineteenth century in Leeds and Manchester for the fire insurance profession by the chemist Herbert Ingle (1860–1945). In 1900, Ingle, who taught agricultural chemistry at the Yorkshire College (the University of Leeds from 1904) combined forces with his younger brother Harry Ingle (1869–1921), who had taken his doctorate in chemistry at the University of Munich before practising as a linoleum chemist in Kirkcaldy, to write a simple textbook of fire hazard chemistry. Like Harris's books, the Ingles' *Chemistry of Fire and Fire Prevention* was advertised as a handbook for "insurance surveyors, works' managers, and all interested in fire risks and their diminution". The textbook steered a curious course between being an elementary chemistry textbook and an account of various manufacturing processes and their dangers.

Theatres were particularly dangerous types of building, given their wooden and canvas sets and their use of arc lighting to illuminate the stage. The Anglo-German architect and engineer Edwin Otho Sachs (1870–1919) dedicated his career to the prevention of theatre fires. Born in London, but educated in Germany, Sachs worked part-time as a volunteer fire-fighter in Berlin, Vienna, Paris and London while engaged in an architectural practice which he summarized in his definitive treatise, *Modern Opera Houses and Theatres* (three volumes, 1896–1898). In 1897, he founded the British Fire Prevention Committee. Within three years this committee had set up a Fire Testing Laboratory which, among other things, tested fireproofing agents and the several new sprinkling devices that had come onto the market. Most of these fire-extinguishers (then called "extincteurs") were of American origin and consisted of a cylinder containing sodium bicarbonate, together with a sealed tube containing sulfuric acid. These carbon dioxide extinguishers became ubiquitous in offices, factories and schools during the following century. Sachs worked tirelessly for the Committee, publishing pamphlets and the booklet *Facts on Fire Prevention* in 1902.

By then, Sachs had become convinced that the answer to theatre fires lay in the use of a construction system that used reinforced concrete for fireproof protection, a system pioneered by the French architect François Hennebique (1842–1921) in the 1890s. In 1906, Sachs founded the bimonthly journal, *Concrete and Constructional Engineering*, to promote knowledge of reinforced concrete among architects and builders. Two years later, in 1908, he was the driving force behind the foundation of a professional organization, the Concrete Institute. Its membership of architects, engineers, surveyors, chemists and manufacturers formed a powerful lobby group to make the London County Council adopt reinforced concrete as part of its Building Acts. Once these political ambitions were satisfied, the Concrete Institute widened its scope to include all aspects of structural engineering, and in 1922 it became the Institute of Structural Engineers, thus disguising its origins in fire prevention. Although a sick man, when the Firemen's Trade Union (later the National Fire Brigades' Union) was founded in 1918, Sachs was made its vice-president in honour of his services to fire prevention.

By the 1920s fire-fighting and the prevention of fire had become a profession with its own qualifying association, the Institute of Fire Engineers, which had been founded in 1918. One of its earliest members was the Scots chemical consultant, Alec Munro Cameron (1886–1965), whose business was conducted from Lasswade. A member of the Institute of Chemistry, he had written a pamphlet on *Fire Risks in Industry* for its members in 1927. In 1933 he was commissioned to write *Chemistry in Relation to Fire Risk and Fire Extinction* for those studying for examinations set by the Institute of Fire Engineers and the Chartered Insurance Institute. Textbooks recommended previously by both Institutes had been too elaborate and advanced; Cameron's book instead concentrated on the chemical reactions involved in combustion and in fire extinction.

Cameron skilfully arranged his account around classes of substances, gases, liquids and solids which might prove hazardous, before examining specific industrial processes that were known to need careful control. These included nitration, distillation, drying and storage in dry conditions. Dry cleaning involving the use of benzol was highlighted as particularly dangerous. The chemistry involved was described in fairly simple terms on a need-to-know basis. Thus, when discussing liquids, Cameron stressed how the insurer needed to know about flash points and how they were determined. Cameron's book, which was a distinct improvement on the one written earlier by the Ingle brothers, went into an enlarged edition in 1948. However, by then insurance companies had begun to employ graduate chemists to assess fire risks and such textbooks became redundant.

Although insurance schemes against domestic fires or loss of merchant shipping at sea had long been the mainstay of the insurance companies that emerged in the seventeenth century, as the Ingle brothers and Cameron noted, the expansion of the chemical industry gave rise to new hazards. The Association of British Chemical Manufacturers, which was founded in 1916, paid particular attention to safety in chemical plants. Its "model rules" for chemical works and its quarterly summaries of safety information did much to reduce the risk of chemical hazards during the interwar years. In the late twentieth century international signing systems such as HAZCHEM began to be widely used, the various codes indicating how to, or not to, treat chemical spillages during transport or storage. Initially the idea of the London Fire Brigade on a voluntary basis, the HAZCHEM scheme for the transport and storage of chemicals became mandatory in 1979.

REACH, the European Union's regulation on the use of chemicals was adopted in 2006 after nearly a decade's discussion (REACH stands for **R**egistration, **E**valuation, **A**uthorization and **R**estriction of **Ch**emical Substances). Its aim is to improve human health and the environment through a universally agreed evaluation of the dangerous properties of all chemical substances whether they are naturally occurring or synthetic materials. The aim is to provide knowledge of chemical properties that makes their handling and storage safe; information is held in a central database by the European Chemicals Agency. All chemical manufacturers, importers of chemicals as well as users of chemicals in secondary manufacturing processes within the European Union must demonstrate that a particular chemical can be used safely for a specific purpose and what risks are involved to employees, handlers, distributors and users. To date REACH consists of over 800 pages of rules and thousands of pages of guidance. How safety officers will cope with such regulations remains to be seen. We have certainly come a long way from William Harris's simple dictionary of chemical hazards for insurance agents.

CHAPTER 34

Math for Chemists

The chemist Alexander Crum Brown, who was descended from a dynasty of distinguished Scottish divines, believed that dynamics and mathematics lay at the heart of the universe. Outside chemistry he spent many happy hours building three-dimensional models of algebraic functions, several of which are now in the London Science Museum. Writing in the *Journal of the Chemical Society* in 1892, he declared his passionate belief that "unless the chemist learns the language of mathematics, he will become a provincial, and the higher branches of chemical work that require reason as well as skill, will gradually pass out of his hands" (Figure 34.1).

In recent years there has been a spate of books published in response to a reputed declining knowledge of mathematics among university and college students. The perennial debate over whether the maths necessary for chemists to know should be taught by chemists or by mathematicians remains unresolved. But when did chemists first begin to teach or demand mathematics—or what American educators call "math for chemists"?

One answer would be found in Jeremiah Benjamin Richter's rare and obscure four volume book, *Anfangsgründe der Stöchyometrie, oder Messkunst chemischer Elemente* (Breslau, 1792–1794). The title translates as First Principles of Stoichiometry, or the Art of Measuring Chemical Elements. The book opens with thirty pages of mathematical instructions on the rules of arithmetic, basic

The Case of the Poisonous Socks: Tales from Chemistry
By William H. Brock
© William H. Brock, 2011
Published by the Royal Society of Chemistry, www.rsc.org

Figure 34.1 Alexander Crum Brown (1838–1922), professor of chemistry at the University of Edinburgh, where he pioneered the use of graphic formulae. Etching by W. B. Hole, 1884. (Science & Technology Picture Library)

algebra, and an account of arithmetical and geometrical series. However, apart from donating the word "stoichiometry" to the field of chemical equivalents and proportions, Richter was not the catalyst who encouraged chemists to learn advanced mathematics. This came about because of the development of physical chemistry in the mid-nineteenth century, and particularly from the need to understand thermodynamics.

The need goes back to the early days of physical chemistry. Walther Nernst (1864–1941), the 1920 Nobel Prizewinner, was never in any doubt that the spirit and methods of mathematics were indispensable in physical chemistry. He collaborated on the very first text with the Göttingen crystallographer, Artur Moritz Schoenflies (1853–1928). Schoenflies is best known as the geometer who postulated the 230 crystallographic groups in *Krystallsysteme und Krystallstruktur* (Leipzig, 1891). Nernst's and Schoenflies's *Einführung in die mathematische Behandlung der Naturwissenschaften* [Introduction to the Mathematical Treatment of the Sciences] was published in 1895 and had gone through eleven editions by 1931. It proved a model for other writers. In fact, as Nernst confessed in dedicating the book to Wilhelm Ostwald, most of the book (which was solely concerned with the differential and integral calculus) had been composed by Schoenflies; Nernst had merely added some chemical applications (Figure 34.2).

The first English "math for chemists" text was by John William Mellor (1869–1938) who had worked through Nernst's

HIGHER MATHEMATICS

FOR

STUDENTS OF CHEMISTRY AND
PHYSICS

WITH SPECIAL REFERENCE TO PRACTICAL WORK

BY

J. W. MELLOR, D.Sc., F.R.S.

NEW IMPRESSION

LONGMANS, GREEN AND CO.
LONDON • NEW YORK • TORONTO
1931

Figure 34.2   Title page from J. W. Mellor, *Higher Mathematics for Students of
Chemistry and Physics*. (Brock)

thermodynamics as a university student in New Zealand before
undertaking graduate studies at the University of Manchester.
Encouraged by Manchester's professor of chemistry, H. B. Dixon
(1852–1930), Mellor published *Higher Mathematics for Students of
Chemistry and Physics* in 1902. By 1919 there had been five edi-
tions. Mellor also prefaced his important *Chemical Statics and
Dynamics* (1904) with a thirty-page mathematical introduction.
Oddly, another Mancunian student, who was certainly aware of
Mellor's successful text, published a rival *Higher Mathematics for
Chemical Students* in 1911. This was none other than James Riddick Partington (1886–1965), the future distinguished historian of
chemistry, who studied with Nernst in Berlin from 1911 to 1913.
Partington closely modelled himself on Nernst, to the extent of
also writing an influential *Treatise on Thermodynamics* (1913)
which remained a standard text for British university students until
the 1950s. Partington's thermodynamics demanded even more
knowledge of the calculus than Mellor's *Dynamics* had done.
Much of Partington's mathematics was later incorporated into his
monumental *Advanced Treatise of Physical Chemistry* (1949–1954)

which, as might be expected, contains a fine bibliography of general treatises on the applications of higher mathematics to physics and chemistry (Figure 34.3).

Both Mellor's and Partington's "math for chemists" are interesting for their use of concrete examples drawn from the chemical literature. The encyclopaedic knowledge of the literature, for which both men became renowned, as well as their deep historical sympathies, is continually apparent. Both men took calculus and determinants to degree standard, since this was the level demanded by University of London BSc examinations right up to the 1960s.

In America, however, Farrington Daniels (1889–1972) at the University of Wisconsin found that both Mellor's and Partington's texts intimidated students, leading him to develop his outstanding *Mathematical Preparations for Physical Chemistry* (1928). It was Daniels who successfully urged the American Chemical Society to launch a national campaign for the improvement of mathematics teaching to science students, a factor that stimulated other

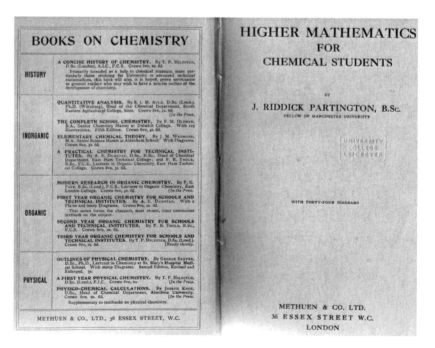

Figure 34.3 Title page of J. R. Partington's rival *Higher Mathematics for Chemical Students*. (Brock)

textbook writers on both sides of the Atlantic. For example, the nuclear physicist Henry Margenau and the nuclear chemist G. M. Murphy collaborated on *Mathematics of Physics and Chemistry* (New York, 1943).

The post-war introduction of quantum chemistry, crystallography, spectroscopy and other topics into undergraduate syllabuses made further formidable demands on students and their instructors. Texts were now frequently written by mathematicians, such as G. J. Kynch, whose *Mathematics for the Chemist* (1955) was aided by Kynch's colleague, the Fourier-transform infrared spectroscopist, Mansel Davies (1913–1995). One Imperial College mathematician, Geoffrey Stephenson, who published a very successful *Mathematical Methods for Science Students* in 1961, remarkably avoided any mention of physical or chemical problems. In practice, the format established by Nernst, Schoenflies, Mellor and Partington has changed very little in the past 120 years and modern authors continue to give credit to the presentational ideas of all these pioneers.

However, with the development of computational chemistry at the end of the twentieth century and with the increasing use of computerized modelling of chemical reactions, it could be said that a mathematical chemistry (analogous to mathematical physics) has emerged. This new research field demands mathematical skills of a high order, as well as a deep understanding of chemistry. The subject was recognized by its own periodical, *Journal of Mathematical Chemistry* in 1975, and by the formation of an International Academy of Mathematical Chemistry in 2005.

CHAPTER 35

The Chemistry of Pottery

Joseph William Mellor (1869–1938) is chiefly remembered and still consulted by chemists for the sixteen volumes of *A Comprehensive Treatise on Inorganic and Theoretical Chemistry* which he composed single-handedly between 1922 and 1937. Its scope and depth were daunting, and many readers and reference chasers must have wondered what sort of library this remarkable chemist possessed. These thoughts are further stirred when confronted by Mellor's *A Treatise on Quantitative Analysis with Special Reference to the Analysis of Clays, Silicates, and Related Materials, being Volume 1 of a Treatise on the Ceramic Industries* (1913). This huge, 778-page illustrated volume reminded me that Mellor, while chiefly remembered for his subsequent and even more monumental volumes, was for the better part of his life a ceramics chemist (Figure 35.1).

Mellor's family emigrated from Yorkshire to New Zealand when he was ten. He left school at fourteen and took evening classes in mathematics, physics and chemistry, as well as modern languages, at Dunedin Technical School while working in a boot factory during the day. Through unremitting hard work he gained a place at the University of Otago in 1892. He returned to England in 1898, newly married, on an 1851 Exhibition Scholarship, gained as a result of a first class degree in chemistry. The scholarship took him to Owens College, part of the University of Manchester, where he began organic research with William Henry Perkin Jr and work on physical chemistry with H. R. Dixon.

The Case of the Poisonous Socks: Tales from Chemistry
By William H. Brock
© William H. Brock, 2011
Published by the Royal Society of Chemistry, www.rsc.org

NO. 48.—THE PIRATED RAILWAY SEAT

Figure 35.1   Joseph William Mellor (1869–1938), ceramics chemist. In this self-
caricature from *Uncle Joe's Nonsense* (1938), Mellor (seated right)
challenges a fellow traveller who had taken his seat in the train.
(Brock)

Mellor had been well trained in thermodynamics at the University of Otago (his immaculately kept lecture notes survive) and
his first book was the pioneering *Higher Mathematics for Students
of Chemistry and Physics* (1902), followed by an important and
influential textbook, *Chemical Statics and Dynamics* (1904). In
1904, he and his wife having decided to stay in Britain, Mellor took
a position as a chemistry teacher (and later Principal) of the newly
formed Staffordshire Pottery School at Stoke-on-Trent. When the
school was absorbed by the British Refractories Research Association in 1915, Mellor became its research director. Running a
small laboratory in the grounds of the North Staffordshire Technical College (now part of Staffordshire University), Mellor
researched the chemistry of glazes and served as official advisor to
the pottery industry. He became a close friend of the eminent
potter, Bernard Moore, and served as Secretary of the Ceramic
Society from 1905 until his death. (The Ceramic Society is now part
of the Institute of Materials, Minerals and Mining.)

Mellor's *Treatise on the Ceramics Industries* is a compendium of advice and instruction for industry. For example, he tells his readers how to determine whether raw materials correspond with the seller's description and suggests questions useful for making informed decisions. Was a given material suited for the purpose required? Was it worth the price? Based upon a course of lectures, the book moves from general analytical procedures and the analysis of simple silicates to procedures for analysing more complex silicates, glasses, enamels and colours; it concludes with advanced techniques for assessing rare elements and other special cases.

Despite his own previous work on physical chemistry, Mellor expressed doubts as to whether the analyst needs to know about "the ionic hypothesis" though he conceded that physical chemistry had made valuable contributions to the theory of absorption, the colloidal state and equilibria. Rigorously documented, the treatise handsomely acknowledged W. E. Hildebrand's pioneering work on silicate analysis for the U.S. Geological Survey. Unfortunately, the promised second volume on more technical aspects of ceramic analysis never appeared, though a second revised edition of the *Quantitative Analysis* was published in 1938, a few months after Mellor's death.

So what happened to his personal library? When Mellor retired from the ceramics industry in 1934, he began to dispose of his enormous collection of books and offprints. Later his widow continued this dispersal, presenting many manuscripts, books and pieces of pottery to the University of Otago. In 1942 his old research laboratory acquired some 25,000 offprints in English, French, German, Italian and Spanish that Mellor had collected since 1898, works that had formed the literary foundation of the *Comprehensive Treatise*, as well as his several textbooks of chemistry. Now based at CERAM Research, the trading name of British Ceramic Research Limited at Stoke-on-Trent, the collection has been greatly augmented by Mellor's successors. The library also contains some of Mellor's undergraduate lecture notes with their beautiful pencil drawings of demonstration apparatus.

Mellor was renowned for his great sense of humour and his ability to lampoon himself in beautifully executed pen-and-ink cartoons. In 1934 his New Zealand relations and his friends in the Ceramics Society persuaded him to collect together the stories of his travels abroad and anecdotes of his meetings with various people. The resulting medley of "fun and philosophy" was published by the Ceramics Society as *Uncle Joe's Nonsense for Young and Old*. Accompanied by his quirky cartoons, the book is unique amongst the publications of professional chemists.

However, Mellor was not unique in building up a large reference collection in ceramics. Another huge library was assembled by the ceramics chemist, consultant and Czech–German refugee, Felix Gustav Singer (d. 1957). Singer, together with his daughter Sonja, published the 1,455-page *Industrial Ceramics* in 1963. Curiously, the Singers made no reference to Mellor's *Quantitative Analysis* in their section on the analysis of ceramics. They did, however, note Mellor's outstanding comment on the physical chemistry of glazing:

"The reaction between the different constituents of the body, in firing, is arrested before the system is in a state of equilibrium. The chemistry of pottery is therefore largely a chemistry of incomplete reactions, and many erroneous inferences have been made from the failure to appreciate this fact. Thus, two bodies may have the same ultimate composition, and yet behave very differently in firing, because the velocity of the reaction, under different conditions, is different when the bodies are compounded with different materials so as to furnish the same ultimate composition, or with the same material in a different state of subdivision."

Clearly, whatever he thought of the ionic theory, Mellor knew that thermodynamics and kinetics were of profound importance in pottery chemistry.

CHAPTER 36

Baker's Dozens

Just as one of the greatest honours conferred by the Royal Society is to be asked to deliver the Bakerian Lecture endowed by the microscopist Henry Baker in 1763, so one of the highest international honours for a chemist is to be awarded the George Fisher Baker Non-Resident Lectureship in Chemistry at Cornell University. G. F. Baker (1840–1931), a founder chairman of the First National Bank of New York, was trained neither as a chemist nor at Cornell, yet from 1912 until his death he gave much of his vast wealth to Cornell as well as to Harvard University.

Following the destruction of Cornell's first chemistry building by fire in 1916, the spectroscopist and analytical chemist Louis M. Dennis (1863–1936), chairman of Cornell's chemistry department from 1903 to 1933, designed the palatial new laboratories whose $1.5 million costs were borne by Baker. The Baker Laboratory of Chemistry, which opened in 1923, was considered to be one of the largest and best-equipped chemistry laboratories in the world (Figure 36.1).

On 14 November 1925 Dennis announced a further coup, an anonymously funded lectureship for:

"the benefit and advancement of teaching and research in chemistry ... Distinguished men of this and other countries and allied fields of science are to be invited to spend one or two semesters at Cornell delivering lectures, conducting research,

The Case of the Poisonous Socks: Tales from Chemistry
By William H. Brock
© William H. Brock, 2011
Published by the Royal Society of Chemistry, www.rsc.org

Figure 36.1 Photograph of the Baker Laboratory building at Cornell University taken by the biochemist George Barger (1878–1939) in 1928 when he delivered the Baker Lectures, *Organic Chemistry in Biology and Medicine*. The photograph was developed using adrenaline, whose synthesis he had attempted in 1909. (Brock)

and generally collaborating with the department while in residence here."

This idea was not entirely new. When Cornell was founded in 1865, its President Andrew D. White had supplemented his small number of teaching staff by inviting famous scholars as non-resident lecturers for short periods to boost the university's instruction, research and cultural life. Visiting celebrities had included the zoologist and geologist Louis Agassiz; the British historian Goldwin Smith, who had supported the reform of science teaching at Oxford; and the poet James Russell Lowell.

The anonymous donor was, of course, Baker, who shared Dennis's feeling that science, like music, was an international language in which differences of race, creed and politics could be forgotten. The lecturers would show the League of Nations how to cooperate. Yet the first lecturer, J. H. van't Hoff's pupil Ernst Julius Cohen of Utrecht, who lectured on the problem of understanding how definite chemical substances could exist in a number of different physical states (*Physico-Chemical Metamorphoses and some Problems in Piezochemistry*, 1928), was to perish in Auschwitz in 1944.

The endowment paid for two lecturers each year—one in the spring and the other in the autumnal semester. A well-equipped laboratory and offices were made available to visitors, though this accommodation gradually went into general use when it was found that visitors generally preferred to collaborate with Cornell chemists using their own equipment. Although Dennis had intended that distinguished scientists in fields allied to chemistry should be sometimes invited, this has only occurred on two occasions. The first was in 1927 when Archibald Vivian Hill, who had won the Nobel Prize for Medicine in 1922 for his work on muscle contraction, lectured on physiology. His *Muscular Movements in Man*, which appeared the same year, contained data on Hill's experiments on the carbon dioxide outputs of Cornellian athletes as they raced around the sports arena. Two years later, the British physicist, George P. Thomson, discussed his work on electron diffraction (*The Wave Mechanics of Free Electrons*) and introduced Cornell's chemists to quantum mechanics for the first time.

Until 1934 all the lecturers adhered to the stipulation to publish their lectures. However, in that year, both Gilbert Newton Lewis (1875–1946), the founder of ideas on electronic bonding and famed for his textbook on thermodynamics, refused to print his lectures on the topic of the day, heavy hydrogen. This appears to have been because Lewis was wary of giving away data to his competitors Harold Urey at Columbia and Harold Taylor at Princeton. Lewis was (rightly) frustrated that he had not been awarded the Nobel Prize for his work on valence and perhaps hoped for greater success with deuterium. Ironically, it was Urey who was awarded the prize for the discovery of deuterium that year. Even ruder, in the same year, the Dutch colloid chemist Johann Rudolf Katz (1880–1938) declined to give any lectures when he arrived on campus. This ungracious behaviour was never repeated, though Lewis's and Katz's failure to produce monographs set an embarrassing precedent. With the exception of Pauling's lectures on chemical bonding in 1937, there were no other published lectures between 1936 and 1939. When the lectures resumed in 1948, publication became the exception, rather than the rule, which is a great pity for the historian of twentieth-century chemistry.

Those lecturers who did publish were often slow in delivering their manuscripts. George Barger's *Some Applications of Organic Chemistry to Biology and Medicine*, given in the spring of 1928, was not published until 1930. William Lawrence Bragg, who based his 1933–1934 lectures on a book already in press (*The Crystalline State*), wrote a book especially for the series, *Atomic Structure of Minerals*, which appeared in 1937.

The first Americans to be awarded the lectureship were the kineticist Farrington Daniels (1889–1972) in 1935 (*Chemical Kinetics*, 1938), the agricultural chemist and biochemist Ross Gortner (1885–1942) in 1936 (*Selected Topics in Colloid Chemistry*, 1937) and Linus Pauling in 1937. Pauling's *The Nature of the Chemical Bond* (1939), which became the most cited of all twentieth-century chemical texts, was not the first or last of the Baker series to achieve textbook status. Nevil Sidgwick's *Some Physical Properties of the Covalent Link in Chemistry* (1933) was adopted as a required text at the University of Sydney and greatly influenced the growth of coordination chemistry in Australia, while Christopher Kelk Ingold's *Structure and Mechanism in Organic Chemistry* (1953) was widely adopted as an undergraduate text in Britain. Other monographs, like Alfred Stock's *Researches on the Hydrides of Boron and Silicon* (1933) with its details of vacuum-line methods for handling volatile substances that decomposed in the presence of air, became influential research volumes. It was his fiancée's gift of Stock's book as a graduation present that prompted Herbert Brown's entry into organic boron chemistry, as he revealed when giving his own 1969 Baker lectures nearly forty years later (*Boranes in Organic Chemistry*, 1972).

To have been asked to give two series of Baker Lectures is an exceptional mark of honour. This has only occurred once. Some unpublished lectures on the determination of molecular structures by X-ray analysis were given in the autumn of 1939 by the Dutch physical chemist Peter Debye (1884–1966), who had won the Nobel Prize three years before. He was on leave from the new Kaiser-Wilhelm-Institute for Physics in Berlin, where the Nazi authorities had recently barred him from the laboratories until he agreed to take out German citizenship. Unwilling to be blackmailed for political purposes, Debye happily decided to stay in America when Cornell offered him a chair of physical chemistry and chairmanship of the department—positions he held with distinction until his formal retirement in 1950. In 1961 (aged 77) he returned to deliver a second course of lectures on molecular forces. Another lecturer recruited to the staff following his lectures was Paul J. Flory (1910–1985), whose talks on polymer chemistry in 1948 (*Principles of*

Polymer Chemistry) were followed by immediate election to a post in the subject the same year.

Like the Nobel lectures, the blue-bound monographs of the George Baker lecturers, which now number over thirty volumes, offer a synopsis of the history of chemistry in the twentieth and early twenty-first centuries, as well as insights into the minds of leading thinkers and experimentalists. They have preserved and encouraged a form of writing, the book, which periodical-loving chemists have largely eschewed since the middle of the nineteenth century.

It is much to be hoped that the series which, in addition to those cited, includes monographs by Ronald Bell (*The Proton in Chemistry*, 1959), Kasimir Fajans (*Radio-elements and Isotopes*, 1931), Otto Hahn (*Applied Radiochemistry*, 1936), Gerhard Herzberg (*The Spectra and Structures of Simple Free Radicals*, 1971), Friedrich Paneth (*Radio Elements as Indicators*, 1928), John Monteath Robertson (*Organic Crystals*, 1953) and Paul Walden (*Salts, Acids and Bases*, 1929) will be the subject of a more serious study in the future. Apart from the war years (1940–1947), the Baker Lectureships have been continuous since 1926.

Part 6: Lost to Chemistry

There have been many people trained in chemistry, including those of us trained in the 1950s and 1960s, who have turned aside from chemistry itself to further their careers in different fields. Some like Edward Frankland's son (another Edward Frankland), Alfred W. Stewart (alias J. J. Conington) or C. P. Snow have turned to literature and novel writing, thus underlying the close relationship between science and the humanities and the role that imagination plays in both. Others (like Anthony Cuthbert Baines, who read chemistry at Oxford) have become distinguished musicians. Others like Aleksandr Borodin (1833–1887), the Czech sugar chemist Emil Votoćek (1862–1950), and the French rare earths chemist George Urbain (1872–1938) have remained tied to the laboratory bench and tried to be part-time composers. Borodin's career has led to perennial debate. Was he a great composer because he was a poor chemist or a limited composer because he devoted too much effort to his chemistry? Edward Elgar, it will be recalled, took a keen interest in chemistry to the extent of even patenting a device for producing hydrogen sulfide, but no one has dared suggest that his music was diminished by this interest. Is it possible to be a part-time chemist? Probably not if the intention is to forge ahead at the frontiers of research. But it is certainly possible to earn a living as a competent chemist and to become well-known for something quite different.

A good example is William Stanley Jevons (1835–1882) who abandoned a promising career in chemistry for one in economics. Nevertheless, he retained a strong interest in scientific developments and became a wise critic of scientific method. Doubtless many people trained initially as chemists have become artists, judges, soldiers or even, if Henry Roscoe's autobiography is to be believed, criminals. And whereas a century ago the vast majority of chemical graduates went into industry, today in the twenty-first century, it is far more common to make a living in banking, accountancy or the law.

What connects the stories in this final section is the way that chemistry provides a liberal education, allowing its exponents to move easily outside the professional discipline into other fields of endeavour. Chemistry's loss can be society's gain. The first vignette concerns a pugnacious nineteenth-century chemist whose part-time passion for violin-making and playing largely destroyed his occupation as a chemical analyst. Music also plays a role in the second tale of a case of mistaken identity, a tale that also incidentally exposes the controversies that have always reigned over Britain's examination culture. In more jocular vein, the third story may suggest that chemistry can drive a person mad and lead away from the laboratory bench to the lunatic asylum. It can often come as a surprise to find that a well-known personality in the arts was initially destined to be a chemist. The story of the graphic artist, humorist and novelist George Du Maurier reveals someone whose talents were better suited outside the boundaries of science.

In the days when universities had the right to appointment a member to sit in Parliament, chemist MPs were not unknown. The most distinguished of these was probably Lyon Playfair (1818–1898), the Professor of Chemistry at the University of Edinburgh from 1858 to 1868 following which he became MP for the Universities of Edinburgh and St Andrews until 1885. When he finally retired to the House of Lords in 1892 he was rightly considered the "elder statesman of science". Playfair never hid his scientific credentials but used them to contribute to political debates. The story of Stafford Cripps in the twentieth century is different. Although trained as a chemist and proud to have been one, he does not seem to have used his scientific knowledge to advance the scientific cause in the way that Playfair did. The final story of C. P. Snow reveals someone who was torn between a career as a spectroscopist or scientific administrator and that of a full-time novelist and writer. That he had the capacity to fulfil all of these roles competently may account for Snow's air of omniscience that many found unattractive.

CHAPTER 37

They Also Ran

It can be interesting to historians of science to know when scientists failed to get university appointments.

At the University of Cambridge there were very few elections before the twentieth century to the chemistry chair that was founded in 1702. The first election did not take place until 1773 when Isaac Pennington narrowly defeated William Hodson by twenty votes. Although the young John Herschel thought of standing when Smithson Tennant died unexpectedly in post in 1815, he deferred to the older James Cumming, who was therefore elected unopposed. However, when Cumming died in 1861 there were three candidates for the chair: George Downing Liveing (who won the contest), B. W. Gibsone and "Dr Phippson". The latter had announced himself as a candidate to the Vice Chancellor, but not having a Cambridge degree, the Registrary declared him ineligible.

"Dr Phippson" was, in fact, Thomas Lamb Phipson (1893–1908), a London consultant chemist and owner of the venerable Godfrey pharmacy in the city of London. His family had emigrated from Birmingham to Brussels in 1849, and Phipson had graduated from the University of Brussels in 1855. Bilingual in English and French, he had spent some years in Paris teaching analysis and helping to edit the weekly science journal *Cosmos* before setting up business in London at the beginning of 1861. An argumentative man, he was a prolific writer on analytical subjects in *Chemical News* and he published many books, the most important being

The Case of the Poisonous Socks: Tales from Chemistry
By William H. Brock
© William H. Brock, 2011
Published by the Royal Society of Chemistry, www.rsc.org

Phosphorescence (1862), which is still consulted. An amateur violinist of distinction, he more or less abandoned science in the 1880s to devote the remainder of his life to music-making and to writing four books on violin-making and famous violinists.

The other mystery candidate in 1861 was certainly eligible. He was the Reverend Burford Waring Gibsone (1828–1896), who had matriculated at Trinity College (Cumming's college) in 1849. He was ninth wrangler in 1853 (Liveing was eleventh wrangler in 1850), having also taken his BA at King's College London in 1851. Like Liveing, Gibsone stayed on at Cambridge to take the then postgraduate Natural Science Tripos in 1854. Again, like Liveing three years before, he was awarded first class honours, with a distinction in chemistry. Unlike Liveing, however, he took holy orders in 1854. Following a brief curacy in Norwood, Surrey, Gibsone moved to Guernsey as Vice-principal of Elizabeth College. It seemed, however, that he still hoped to pursue a scientific career. Between 1855 and 1857 he held the Chair of Mathematics at Queen's College, Birmingham, one of the several institutions that eventually united to form the University of Birmingham. There he embarked upon an external London science degree, obtaining his BSc with first class honours in 1861. By then he had founded his own school, Grosvenor College, in Bath and it was from there that he applied for the Cambridge chair. The electors ignored him completely, and all 26 voters plumped for Liveing who had the advantage of a Cambridge base and of already having been Cumming's deputy. Times (and salaries) had also changed since 1815, and the idea of a clergyman don (as Cumming had been) who would spend half the year away from Cambridge performing parish duties was no longer acceptable.

Gibsone did not pursue the science track further. He moved to the Mercers' School in London as deputy headmaster before completing his educational career from 1866 as head of St Peter's College in Easton Square, London. In the 1870s he returned full-time to a religious avocation, ending his life as Vicar of Wolvey in Warwickshire, where he has a memorial. His sole contribution to chemistry was the addition of a safety device to prevent the escape of hydrogen sulfide from Kipp's apparatus in 1864.

The winning candidate, George Liveing (1827–1924), proved a conscientious teacher in the tradition of Cambridge's tradition of liberal education. He had already established a laboratory at his own expense at St John's College and encouraged other Cambridge colleges to do the same. An astute academic politician, in the 1890s he succeeded in persuading the university to erect a well-equipped chemistry laboratory in Pembroke Street. His interest in spectroscopy led to an uneasy research collaboration with James Dewar, who held the Jacksonian chair of natural philosophy as well the direction of the Royal Institution in London. He retired in 1908 to be succeeded by a complete outsider, the research-active Londoner, William Jackson Pope, a pupil of Henry Armstrong's. Liveing was expected to become a centenarian chemist like Michel Chevreul (1786–1889) and in his long active retirement he became the butt of jokes on the lines of whether "living was dead". Sadly, he died in December 1924, two months after he was knocked down by a lady undergraduate from Girton College as he walked to the laboratory.

CHAPTER 38

Who Was Crookes's Musician–Chemist?

William Crookes published an annual list of chemistry courses for students every September in his *Chemical News* for nearly forty years from 1863 until 1900. These listings were always prefaced with an editorial commentary on the current state of chemical education in Great Britain. In one such editorial in September 1879 Crookes commented on an address that T. H. Huxley ("that Ithuriel of the 'Cram Demon'"; the reference is to an angel in Milton's *Paradise Lost*) had given at a Speech Day at University College School the previous July. Huxley had delineated the characteristics of prize-winners as quickness in learning, readiness and accuracy in reproducing learned materials, industry and endurance. Crookes' jaundiced comment on this was that Huxley's ideal was rarely found in practice. More usually, glittering prizes encouraged cramming. He recalled a student who had never performed a dissection in his life carrying off a prize in the London University BA exams by gaining honours in animal physiology by dint of possessing a good memory. Crookes, who admired a book entitled *The Boyhood of Great Men* (1853 and later editions) by John George Edgar, also thought someone should write a complementary volume entitled *The Manhood of Great Boys* to encourage youths who failed to carry off prizes at school of university.

The Case of the Poisonous Socks: Tales from Chemistry
By William H. Brock
© William H. Brock, 2011
Published by the Royal Society of Chemistry, www.rsc.org

Both Huxley and Crookes regarded competitive examinations as only a partial test of what prize-winners would achieve in later life. The aforementioned student, for example, had ended up as a government civil servant and not as a biologist. It was at this point that Crookes alluded to a contemporary of this student who had carried off all the BA chemistry prizes in the same graduation year "without ever having cleaned a test tube in his life". This particular student was now (in 1879) "one of our leading musical composers and critics".

Who might this have been? It certainly was not Granville Bantock, Sterndale Bennett, Frederick Delius, Edward Elgar, William Grove, William Hadow, Hubert Parry, William Pole, John Stainer or Charles Villiers Stanford—to name some obvious candidates from among late Victorian composers and musicologists.

An intriguing suggested identification is Richard Carte (better known as D'Oyly Carte), the founder of the Savoy Opera Company and the impresario who sponsored the partnership between Gilbert and Sullivan. However, although Carte would have studied chemistry at University College School and began studying for a degree at University College in 1861, he abandoned studies before taking a degree. In any case, Carte never wrote any musical criticism.

It turns out, as some sleuthing in the University of London's *Historical Record* reveals, that Crookes must have thought the former student concerned was Ebenezer Prout (1835–1909), an important (and underrated) Victorian musician who is best remembered today for the Novello edition of Handel's *Messiah* (1902) which most choral societies still use. The problem with Crookes's identification is that Prout never won a BA prize for chemistry. Prout's father (also named Ebenezer) was a Congregational minister who had married a sister of Edward Stallybrass, another Nonconformist minister. In 1837, two years after Ebenezer Prout's birth, the family moved from Oundle to Halstead, Essex where his father became minister at the Old Meeting and, from 1845, Secretary of the London Missionary Society.

Young Prout was educated at the grammar school in Denmark Hill, London, and spent a year at New College in St John's Wood, 1851–1852. This College, founded in 1850 as an offshoot of Homerton College in Hackney, was a training institution for Congregational ministers. He soon decided against studying for the ministry and instead became a teacher at the Priory House School in Clapton from where he took a London "external" degree in 1854. But there was no prize for chemistry. In 1859, finding that he enjoyed teaching singing more than languages or literature, he began a successful career as a largely self-taught pianist, organist, composer, critic and musicologist. By 1879, when Crookes was writing his *Chemical News* editorial, Prout was known as a composer for piano, organ and strings, and for a number of cantatas. Books on instrumentation, harmony and orchestration had established him as a musicologist, and he was also music critic for *Academy* and *Athenaeum*, as well as the founding editor of *Monthly Musical Record*. There can be no doubt, then, that he was the person Crookes had in mind.

Unfortunately, Crookes had confused Ebenezer Prout with his younger brother Edward Stallybrass Prout (1836–1925) who had taken his BA degree in 1855 while also at New College. It was Edward, not Ebenezer, who won the chemistry prize for his finals papers. Whether it was fair to say that he was a "crammer" who had never washed a test tube only an examination of New College's scientific teaching would reveal. In any case, the University of London's external examinations did not involve practical examinations in the 1850s. Edward's talents were lost to chemistry. Nor did he display any of his brother's musical ability. Instead he followed in his father's footsteps by serving as a Congregational minister in Norwich, Doncaster and Bridgwater before joining the British and Foreign Bible Society in 1885. He retired to Reading where he died on 11 November 1925, the author of a couple commentaries on the Gospels.

While it is easy to understand how Crookes confused the two brothers (I myself wondered initially if Prout the musician had been christened Edward and taken the name "Ebenezer" when he became a professional musician), the real puzzle is why a prize-winner's success in 1855 should be recalled by Crookes twenty-four years later in 1879. This is elephantine memory.

The answer is probably twofold. In 1855, when Edward Prout won his prize, Crookes was working as a meteorological observer at the Radcliffe Observatory in Oxford. He was, however, still in close contact with former colleagues and students at the Royal College of Chemistry (RCC) where he had served as one of A. W. Hofmann's assistants from 1850 to 1854. No doubt Edward

Prout's prize for textbook knowledge was the subject of malicious comment about cramming among the practically oriented members of the RCC and that Crookes got to hear about it this way. The surname Prout would have undoubtedly attracted chemists' attention because of the eponymous hypothesis that the so-called elements were polymers of hydrogen. Prout's hypothesis was to play a central role in Crookes' speculations about the nature of matter throughout his career. Although William Prout, a London physician and clinical chemist, had died in 1850, it is also possible that Crookes thought that Edward/Ebenezer Prout was one of his sons. In fact the families were unrelated. Probably, then, it was Crookes's interest in Prout's hypothesis that made him remember the case of a student named Prout who had won a prize for knowledge of chemistry without having studied it experimentally. To Crookes, writing in 1879, "a more practical and instructive commentary on the evils of 'cram' could hardly be found" than was afforded by Prout's prize.

CHAPTER 39

The Chemist from Hanwell Asylum

Friedrich Wöhler's 1840 substitution satire, written ostensibly by S. C. H. Windler, is well known. In it Wöhler ridiculed Dumas's theory of substitution in a spurious report that he had succeeded in replacing all the atoms in manganous acetate by chlorine. Wonder of wonders, the chlorine still retained all the properties of manganous acetate: it had remained true to its type. In contrast, a later chemical satire from a very intelligent "inmate of Hanwell Asylum" has passed unnoticed by historians of chemistry. In July 1864 the supposed lunatic from the Middlesex County Asylum at Hanwell sent a letter to William Crookes's *Chemical News* declaring that he had been driven mad by the way in which chemists were using different formulae for the same substances. Just look at water, the correspondent suggested.

> "One writer formulates water as HO, another as H_2O, another as $H_2\Theta$, a fourth *HO*; in one page we find slaked lime appearing as CaO.HO in the next *CaO.HO*, a few lines further on CaHO, then CaHΘ, then ~~CaH$_2$O$_2$~~, then CaH_2O_2 until the brain first becomes confused, then swims, and finally softens."

He ended an amusing letter by saying he had asked his "keeper" to post the letter for him by giving him sixpence for a glass of brandy and HO, *HO*, H_2O, $H_2\Theta$, HOH, **HO**—or whatever it was.

The Case of the Poisonous Socks: Tales from Chemistry
By William H. Brock
© William H. Brock, 2011
Published by the Royal Society of Chemistry, www.rsc.org

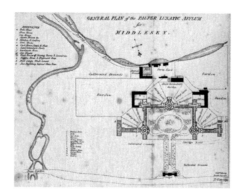

Figure 39.1 A plan of the Hanwell Asylum at Colney, near Southall, which opened in 1831. (Wellcome Images)

Crookes added an editorial note to the effect that it would be a great help to editors if standardization was introduced. To that end, Crookes suggested that an international congress of chemists ought to settle the question—an ironic suggestion given that a meeting to settle this very question had already met at Karlsruhe five years before without resolving the issue. If a "quiet revolution" over molecular standards was nevertheless taking place, as Alan Rocke has argued, the Hanwell chemist was obviously not aware of it (Figure 39.1).

A year later, in October 1865, the Hanwell chemist, who now jokingly signed himself "N. H. Three", published a second letter headed "Water From a Maniacal Point of View". Unable to experiment because his keepers would not allow him access to chemicals, he had been forced to think instead. It occurred to him, he wrote, that ammonia, which A. W. Hofmann had suggested belonged to a model ammonia type in series with the hydrogen, hydrogen chloride and water types, was really a water type. In proof of this, the Hanwell chemist had been able to replace trivalent nitrogen, N''', by univalent hydrogen (H') and bivalent oxygen (O''), and one of the other hydrogen atoms by hydroxyl, so that:

$$
\begin{array}{ccc}
& H' & \\
N''' & H' \quad \text{became} \quad &
\end{array}
\qquad
\begin{array}{c}
HO' \\
H' \ H' \\
O'' \ H'
\end{array}
$$

or what he curiously termed "dihydrohydrorylhydroxamine". This may have been a misprint for "dihydrohydroxylhydroxylamine" using hydroxylamine ($NH_2.OH$) as the base molecule for substitution. In view of this astonishing substitution, he was now hard at work reducing marsh gas to ammonia, for which reason he signed off "yours ammoniacally".

It is interesting to note that Crookes did not introduce structural formulae into *Chemical News* until 1866, when he reviewed Edward Frankland's *Lecture Notes for Chemical Students*. However, they were used sparingly until 1869 and he did not deploy a benzene ring until ten years later.

Who was the Hanwell chemist? He was clearly familiar with Wöhler's earlier joke, even he if made no explicit reference to it, and he was very clearly familiar with what was going on in contemporary organic chemistry. He must have been a trained chemist. Whether he was a genuine inmate of the huge lunatic asylum at Hanwell would require a detailed examination of the asylum's records.

There is, however, another likely possibility. One of the most promising young chemists in the 1860s was Ernest Theophron Chapman. A sickly child from Clapton (whether in Gloucestershire or Somerset is unclear), he had taught himself chemistry by reading W. A. Miller's *Elements of Chemistry*. At the age of fifteen (like A. W. Williamson earlier) he had gone to school in Heidelberg (where he also attended some of Robert Bunsen's lectures at the University). On returning to London he joined the Royal College of Chemistry where he learned about the ammonia type from A. W. Hofmann and the new ideas of valence from his successor Edward Frankland. The latter sent him back to Germany to study at Marburg with his great friend, Hermann Kolbe.

On his return to London, Chapman joined the laboratory of an equally promising chemist, Alfred J. Wanklyn, a former pupil of Frankland's time at Owens College in Manchester. Wanklyn had been made Professor of Chemistry at the London Institution in the City of London where, together with Chapman and Miles H. Smith (whose later career remains a mystery), the three chemists developed a method of water analysis using ammonia. The presence of organic nitrogen (assumed as a contaminant from sewage) was estimated as "albuminoid ammonia" by distilling water samples with alkaline permanganate and then titrating the ammonia formed. The method proved controversial and was heavily criticized by Frankland who, together with Henry Armstrong, developed a rival method that involved a complex gravimetric procedure. Chapman also developed his skills as an organic chemist, publishing a number of distinguished papers that used Frankland's structural notation rather than type formulae. For example, he prepared pyridine by the dehydration of amyl nitrate in 1868, and published a score of papers on the derivatives of organic compounds subjected to slow oxidation procedures. He was clearly a modernizer. Like Wanklyn, however, he was a young Turk and became one of the prominent critics of the way the Chemical Society was being run. During a blackballing episode which nearly split the Chemical Society asunder, Chapman managed to get himself elected to Council as a reformer in 1869. In August of the same year, whether in disgust at the British chemical establishment or because of promising career advancement, he left London to become the manager of an explosives factory in the Hartz Mountains in Saxony. There he blew himself up accidentally on 25 June 1872 while preparing a batch of "methylic nitrate" [ethyl nitrate], commonly used as a dynamite substitute in rock blasting. He was just 26 years of age.

Frankland, who read and probably composed the brief obituary for the Chemical Society, mourned Chapman's death as "a serious deduction from the investigating powers of this country". Chemistry could ill afford "to lose from her ranks an original investigator at once so active and so successful".

Reading between the lines of Frankland's obituary, it is clear that Chapman's health was never robust and that repeated bouts of illness included frequent periods of depression. Chapman constantly suffered from haemorrhaging from the lungs, which he attributed to inhaling chlorine accidentally when working at the Royal College of Chemistry; this is more likely to have been due to tuberculosis which is also associated with periods of elation and depression. Did the fact that he lived at Oak Cottage, Hanwell, close to the huge asylum, appeal to his sense of humour while he was in a dark mood and lead to his two satirical letters?

CHAPTER 40

George Du Maurier (1834–1896)

Readers of David Lodge's novel, *Author, Author*, a fictional study of the Anglo-American novelist Henry James and of his failure to gain recognition as a playwright, will have come across James's friendship with the graphic artist George Du Maurier (1834–1896). A much-admired *Punch* cartoonist, Du Maurier also wrote the best-selling novel *Trilby*. His son, Gerald, became a famous actor and his grand-daughter, Daphne Du Maurier, an equally-famous novelist. It comes as a surprise, however, to find that George Du Maurier had begun his career as an analytical chemist; but this is no fictional invention of Lodge's (Figure 40.1).

The grandly named George Louis Palmella Busson Du Maurier was born in Paris in 1834, the son of Louis-Malthurin Du Maurier, who is aptly described by the new *Oxford Dictionary of National Biography* as "an improvident scientist and inventor", and his English wife, Ellen Clarke. The aristocratic elaboration, Du Maurier, in the family name was the invention of his grandfather, a French glassblower who worked for some years at the glassworks at Whitefriars in London. The grandfather had delusions of grandeur and held the pretence that the family had aristocratic French and Portuguese ancestry. George Du Maurier, who never knew of this fiction, failed his baccalauréat in 1851, in which year the family settled in England in the hope that Du Maurier senior could exploit his invention of a carbide lamp. Probably to help his father's ambitions, in the same year young Du Maurier enrolled in

The Case of the Poisonous Socks: Tales from Chemistry
By William H. Brock
© William H. Brock, 2011
Published by the Royal Society of Chemistry, www.rsc.org

Figure 40.1 George Louis Palmella Busson Du Maurier (1834–1896), who trained as a chemist at University College London before becoming an artist in black and white, and a novelist. (© National Portrait Gallery, London)

Alexander Williamson's chemistry classes at the Birkbeck Laboratory at University College London. Williamson was a highly successful teacher and researcher despite having a withered arm and blindness in one eye. In a jocular letter to one of his aunts, Du Maurier wrote that:

"I am from 9 till 2 in the laboratory, testing all the nastiest substances that were ever enclosed in glass bottles; just fancy me with an old coat, and a stained black apron down to my feet, scarcely visible among fumes and vapours of all colours and smells; in one hand a sandwich, in the other a bottle of sesquifferocyanide [*sic*] of potassium, or protosulphate of iron, or other substances with names no less euphonious. I am full of ardour in the pursuits of my profession; indeed I am quite forgetting the usual terms of common things, and instead of asking for salt, or water, etc., I ask for chloride of sodium or the protoxide of nitrogen [*sic*]."

By his own account he did not enjoy the lab, spending much of his time sketching, singing, and visiting the British Museum and London art galleries. One of his early drawings is a caricature of himself singing to fellow students in the Birkbeck laboratory. He did not take a degree, but left University College in December 1852 to work as an analytical and consultant chemist in a laboratory his father had erected on the top floor of his house in Barge Yard, Bucklersbury, behind the Mansion House in the City of London. The enterprise was not a success.

In a short story, "Recollections of an English goldmine" that Du Maurier published in Dickens's *Once a Week* in 1861, it is clear that one of his very few commissions was to analyse mineral deposits for gold in a Devonshire copper mine. In the 1850s, following the gold rushes in Australia and California, a large number of British companies were formed to exploit the mercury amalgamation process for gold extraction using American machinery. Du Maurier describes how a maternal uncle had introduced him to the chairman of the Victoria Gold and Copper Mine who, to boost the confidence of shareholders, commissioned Du Maurier to confirm the positive findings of two other "expert London chemists". The company had already spent £6,000 without any return of gold over a six months period. In his amusing account he describes how his distillation of some 30 tons of amalgam had merely yielded a minute button of gold worth $1/7\frac{1}{2}$d (about 3p).

Du Maurier recommended that operations should cease immediately, only to be told to do more tests. Eventually, the company's directors, wives and daughters all came down from London to take a look for themselves. Everyone enjoyed a picnic and drank a good deal (the subject of a bucolic sketch), the directors deciding that the enterprise was a flop. Nevertheless, they demanded that Du Maurier should write a "matter of fact" report that would show shareholders that everything that could have been done had been done. However, when Du Maurier produced a too dry and technical report, he was asked to dress it up verbally at a shareholders' meeting. Instead, if Du Maurier is to be believed, in his speech to shareholders he stated explicitly that the mine contained no more gold than seawater and accused "Mr Ex and Mr Zed" (the unknown consultants whose analyses had prompted the £6,000 investment) of "ludicrous incompetence". If Du Maurier really did say this, it would be hard to see how he could have continued his

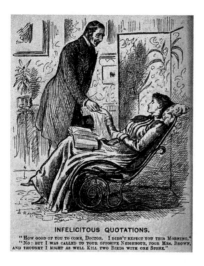

INFELICITOUS QUOTATIONS.

"How good of you to come, Doctor. I didn't expect you this Morning."
"No: but I was called to your opposite Neighbour, poor Mrs. Brown,
and thought I might as well Kill two Birds with one Stone."

Figure 40.2 A doctor visiting a patient, a *Punch* cartoon by Du Maurier.
(Wellcome Images)

career as a London consultant, even if his heart had been in an analytical career. Ironically, the mine was sold to another company, which continued to produce copper profitably for a number of years (Figure 40.2).

This debacle, together with the death of his father in 1856, caused Du Maurier to abandon chemistry once and for all. Instead, he took up painting, training as an artist in Paris, where his bohemian lifestyle was later vividly portrayed in the novels *Peter Ibbotson* (1891) and *Trilby* (1894). Unfortunately, this promising career was cut short in 1857 when a detached retina caused him loss of sight in one eye. As Alexander Williamson's brilliant career as a professor at University College London demonstrates, this disability would not necessarily have prevented Du Maurier from having continued as a chemist, but it was fatal for an artist. Instead, Du Maurier turned to black and white drawing, developing a successful career as a book illustrator and *Punch* cartoonist from the 1860s onwards. When his remaining good eye was threatened in the 1880s he turned to writing autobiographical novels, dictating them to his wife. *Trilby*, the story of a girl who achieves operatic success under the hypnotic influence of the evil Svengali, brought him international fame. But his dreadful posthumous fantasy novel *The Martian* (1897) is best forgotten.

CHAPTER 41

Sir Stafford Cripps

In a lecture on glass tubing given to the British Association of Chemists in 1946, Richard E. Threlfall commented that it would "be interesting to see to what extent the early training of the President of the Board of Trade in chemistry [might] influence him in his policy towards you and us" in austere post-war Britain. Threlfall was referring to the politician Sir Stafford Cripps (1889–1952). Many older chemists will remember him best as Chancellor of Exchequer between 1947 and 1950 when he enforced extremely tough wage restraints and devalued the pound. Few will know or remember that he had begun his career as a chemist (Figure 41.1).

Richard Stafford Cripps came from the traditional and privileged governing classes and was brought up as a Tory and a devout Christian. Both his father and grandfather had been successful lawyers, and his father was raised to the peerage in 1910 as Baron Parmoor, the name of the house and estate the family owned in the Chiltern Hills. His mother, Theresa Potter, was a sister of the pioneer socialist Beatrice Webb. As a child Cripps enjoyed the freedom of 400 acres of countryside in which he could safely ride horses, dam streams, and ride his own motorized bicycle from the age of 12. By all accounts he not only displayed great intellectual ability as a child, but also acquired and demonstrated manual skills such as carpentry and metalwork. His physical skill and dexterity were remarkably displayed in the full-size glider he constructed

The Case of the Poisonous Socks: Tales from Chemistry
By William H. Brock
© William H. Brock, 2011
Published by the Royal Society of Chemistry, www.rsc.org

Figure 41.1 Richard Stafford Cripps (1889–1952), who after studying chemistry at University College London became a lawyer and politician. In this photograph he is captured as a Labour candidate canvassing a baker in Bristol during the October 1931 election. (Science & Technology Picture Library)

from bicycle parts at Parmoor during a university vacation in 1909. Launched from a hilltop, the glider flew, but crash-landed.

Following education at a couple of preparatory schools, Cripps entered Winchester College in 1901 just at the time when a new Warden (headmaster), Dr Hubert M. Burge, was modernizing the school's curriculum by making Greek non-compulsory and improving the college's science teaching. Cripps, who enjoyed the heady atmosphere of the school, was one of the first Wykehamists to study science seriously and he proved outstanding at chemistry. Traditionally, Wykehamists left the college to matriculate at New College, Oxford. Fortuitously, when Cripps applied to Oxford in 1907 New College was for the first time offering an entrance scholarship in chemistry, even though it had no laboratory and its students had to use the outdated facilities of the Oxford Museum directed by William Odling. The external examiner for this scholarship was none other than Sir William Ramsay who was so impressed by Cripps's scripts that he urged him to enter his own laboratory at University College London. Despite his father's disappointment that his son seemed to be abandoning the prospect of a legal and parliamentary career for a scientific one, Ramsay's observation that laboratories at University College London were better than any at Oxford won the day.

For two years Cripps followed the University of London curriculum for the BSc degree. Ramsay always favoured teaching chemical skills to his brightest students by setting problems. Accordingly, in his final year, Ramsay switched Cripps onto a research track for the MSc degree and appointed him a demonstrator. Research on the rare gases that Ramsay and Morris Travers had specialized in was still intense, and after Travers' departure to Bristol in 1904, Ramsay's chief collaborator was Robert Whytlaw-Gray (1877–1958) whose then interest was in the accurate determination of the specific gravities of the inert gases in order to calculate their precise atomic weights. In order to weigh minute quantities of these gases at known temperatures and pressures, Whytlaw-Gray had constructed a micro-balance sensitive to 2×10^{-9} g. Under Whytlaw-Gray's supervision and with the assistance of the chief demonstrator Hubert Sutton Patterson, Cripps helped determine the critical constants of xenon and its density when in equilibrium with its liquid phase (*i.e.* the orthobaric density).

Ramsay and Travers had measured the density of liquid xenon at its boiling point in 1901 and recorded it as 3.58, giving the gas an atomic volume of 36.6. This made it anomalous since the atomic volume of krypton exceeded it at 37.8. Ramsay's Anglo-German graduate student George Rudorf had determined the atomic volume of xenon in 1909 as 48.8, which seemed more reasonable. From van der Waals equation it also appeared that the ratio of the critical density of xenon to its theoretical density at the critical point was 2.667. However, a range of values between 2.62 and 3.6 had been found experimentally by Rudorf and by Daniel Berthelot in France. Meanwhile, Sydney Young's investigation of polyatomic gases such as CO_2 and SO_2, as well as organic liquids, implied that the ratio should be nearly constant at 3.77 if they were all non-associated substances. An experimentally determined value as low as 2.667 suggested that liquefied xenon might be partially associated. Here were good reasons for Ramsay and Whytlaw-Gray to conclude that fresh experimental evidence was required and that this should be part of a programme to determine the critical constants of xenon.

Ramsay supplied Patterson, Cripps and Whytlaw-Gray with University College London's complete stock of xenon amounting to 120 cubic centimetres (cc) at normal temperature and pressure (NTP). This sample was repeatedly purified by solidification and distillation, the final 20 cc being used for the measurements.

The gas was pumped into a graduated U-tube of 1 mm bore that could be attached to a modified form of the apparatus Thomas Andrews had used in determining critical constants from 1869 onwards. At least four readings of the liquid and vapour volumes of xenon were taken at different temperatures ranging between −66 °C and 16 °C. The work involved intricate checks on purity and accuracy. From the data obtained the orthobaric densities were calculated in grams per cc.

It had been shown by L. Cailletet and E. Mathias in 1886, and verified by Sydney Young at University College London in 1900, that the mean of the densities of substances in the liquid and saturated vapour states was a linear function of temperature, $\delta_t = \delta_0 + \alpha t$, where δ_0 and α are constants for a given substance. This is known as Cailletet's law of the rectilinear diameter. By plotting orthobaric densities at a cluster of temperatures near the critical point, the mean density at the critical temperature can be calculated. From this information the critical volume can be deduced. The resultant mean density from fifteen readings of the density of liquid and saturated vapour was 1.205. This meant that xenon obeyed Cailletet and Mathias's law of rectilinear diameters.

Measurements of xenon's critical temperature and pressure were also made (16.6 °C and 58.2 atmospheres respectively). From this information, and using Cailletet and Mathais's equation, the critical volume was found to be 0.866 cc per gram. The team estimated that the boiling point of xenon was −106.9 °C (they did not use the Kelvin scale) and that its liquid density at this point was 3.063 grams per cc, giving an atomic volume of 42.7. As for the van der Waals ratio of critical and theoretical densities, the team found 3.605; somewhat, but not significantly, lower than Young had found for non-associated liquids. The accuracy of the team's work is unassailable. All their values for the critical constants are either identical, or very close to, modern values. The exceptions are xenon's boiling point, corrected from 166.1 K to 163.9 K, and its atomic value, corrected from 42.7 to 42.92.

The triple-authored paper on xenon's critical constants was not presented to the Royal Society by Ramsay until 21 March 1912. The paper ended with a report on some peculiar effects noticed during the experiments which the team initially put down to oxygen impurities, or to the presence of an oxygen compound. Intricate and time-consuming checks on these possibilities, which must have been done by Patterson, lengthened the investigation and this probably explains why work done by Cripps in 1910 was not published until 1912. The eventual conclusion was that weak molecular association between oxygen and xenon akin to solution could not be ruled out as the cause of fluctuating readings. Given the inert gas electronic structure, it seems more likely that weak van der Waals forces were responsible for inconsistencies between experiments.

According to one Cripps biographer, Simon Burgess, the measurements involved Cripps in designing and building a special pyknometer; however, there is no evidence in the published paper that a pyknometer was used in the investigation. Curiously, Patterson, Cripps and Whytlaw-Gray did not calculate the atomic weight of xenon from their data. It would have been 130.79 compared with the modern value of 131.29, the difference being due to their under-determination of the atomic volume. It is interesting to note that Patterson eventually joined Whytlaw-Gray at the University of Leeds where, as an aside to research on aerosols, they re-determined the atomic weight of xenon in 1931 as 131.26.

During his final year at University College London, Cripps was elected President of the Students' Union and made friends with another research student, Alfred ("Jack") Egerton (1886–1959) who was to marry Cripps's sister Ruth in 1912 and achieve an international reputation for research on gaseous combustion. Sir Alfred Egerton (as he became) was a notable product of Ramsay's research school at University College London. On gaining his MSc in 1910, Cripps spent the summer vacation learning German in Hamburg—a sign, surely, that he intended to spend further time in Germany studying for a doctorate in chemistry. Chemistry certainly still occupied some of his thoughts. In a diary entry for 15 July 1910 he wondered why chemical elements were made to fit into the periodic table.

However, by the time the paper on xenon was presented to the Royal Society in 1912, Cripps (unlike Egerton) had abandoned chemistry for the law. The change of mind had taken place during the autumn of 1910 when he helped his father win the parliamentary seat of Wycombe, Buckinghamshire, for the Conservatives. A fellow young helper was Isobel Swithinbank, whose mother was the daughter of James Crossley Eno, the millionaire pharmaceutical manufacturer of "Eno's Fruit Salts". The young electioneers quickly fell in love and announced their engagement in January 1911. Although Isobel stood to inherit a fortune from her mother and £500 a year on her marriage, Cripps decided that if they were to marry he had to earn a sufficiently high income to support Isobel in the "aristocratic" life style to which she was accustomed. A career in academic or industrial chemistry did not promise huge financial rewards. Accordingly, to Lord Parmoor's relief, Cripps began to read for the Bar and he and Isobel were married in July 1911.

That was not, however, quite the end of Cripps' chemical career. A lifelong digestive problem (eventually diagnosed as colitis) prevented him from joining the armed forces when war broke out in 1914, by which time he was a junior barrister in his father's chambers. Instead he did voluntary work with the Red Cross Ambulance Service in northern France. Meanwhile, Egerton had been recruited by the Ministry of Munitions following the scientific community's campaign for chemists to be recruited to aid the war effort. It was Egerton, supported by the dying Ramsay, who recruited Cripps for training at Waltham Abbey in 1915 to help in the manufacture of guncotton and tetryl for TNT detonators. (Tetryl, otherwise known as nitramine, is trinitrophenyl-methylnitramine.)

After two month's training, Cripps was appointed assistant manager of a new tetryl factory at Queensferry in Cheshire. Here he oversaw some 6,000 workers and worked a seven-day week and some twelve to fifteen hours daily. After working flat out for four months, Cripps collapsed and although he returned again, by September 1917 the severity of his colitis forced his retirement. It was not until January 1919 that his doctors allowed him to return to legal practice, where he specialized in patent and railway disputes, and revised many legal texts. To this day, "Cripps test" is used by lawyers to assess the validity of a patent via its "inventive steps"—namely, had it been obvious to any practising scientist or technologist when a patent was filed that it really could be manufactured or processed.

In September 1930 Cripps was knighted on his appointment as Solicitor General and Labour MP for West Bristol. A glittering parliamentary and political career followed in which he strove to counteract the evils he now saw in capitalism and imperialism. He was British Ambassador in Moscow (1940–1942), Leader of the

House of Commons and Minister of Aircraft Production (1942–1945), Minister of the Board of Trade (1945–1947) and Minister of Economic Affairs and Chancellor of the Exchequer from 1947 until 1950, when his health gave way completely. His proudest moment, he told Egerton, was his election to the Royal Society in 1948 as "a person who had rendered conspicuous service to science" (Statute 12 of Fellowship Rules)—a reminder of the great chemical career that could have been his for the taking.

CHAPTER 42

The Chemist as Novelist:
The Case of C. P. Snow

Although it is well-known that the English novelist Charles Percy
Snow (Lord Snow, as he became in 1964) began his career as a
scientist, critical comment and analysis has usually focused upon
his literary output, including his participation in one of the twentieth century's greatest literary "rows"—the debate with F. R.
Leavis over the "two cultures" in 1959. Snow's well-meant warning about what he saw as a growing gulf between scientists and
literary intellectuals was soon lost in a diatribe of personal abuse
and nonsense that only tended to highlight Snow's original point
that those who proudly flaunted their uninterest and ignorance
of science meant that they were blind to major questions concerning the future of human culture such as the uses of atomic
energy, the threat of over-population, and the gap between rich
and poor people and nations.

The present essay, while offering some further background for an
appraisal of Snow's third novel, *The Search* (1934), and suggesting
some reasons for Snow's decision to abandon scientific research
(though *not* the scientific profession) after 1935, redresses the balance by concentrating primarily upon his career as a physical
chemist. Although Snow preserved his literary papers, the bulk of
which are in The Harry Ransom Humanities Research Center of
the University of Texas at Austin, he seems to have destroyed his

The Case of the Poisonous Socks: Tales from Chemistry
By William H. Brock
© William H. Brock, 2011
Published by the Royal Society of Chemistry, www.rsc.org

laboratory notebooks and scientific correspondence. Analysis has, therefore, to be more or less entirely based upon Snow's published scientific writings.

As Snow himself quickly recognized as a postgraduate student, his research career at the University of Cambridge occurred during one of the most exciting periods of twentieth-century British science when Ernest Rutherford and Gowland Hopkins and their respective schools of physics and biochemistry were exploring the nature of the atom and of life. Snow's own research shared in this excitement, for the spectroscopic problem which he found himself investigating seemed to promise an explanation of an ultimate mystery of chemistry—the deep structure of molecules and the reasons for their reactivity.

FROM THE PROVINCES TO THE MECCA
OF BRITISH SCIENCE

Snow was born in the provincial red-bricked hosiery town of Leicester in 1905, the son of a shoe factory worker, part-time music teacher and church organist. After education in a private elementary school, he attended Alderman Newton Grammar School in the city, where his interest in chemistry was awakened. His formal education complete at sixteen, when he passed his matriculation examinations, he remained at the school for a further three years (1922–1924) as the school's laboratory assistant. This allowed him time to prepare himself as an external student for the tough London Intermediate BSc Examination, and to read widely in English and European literature.

In 1921, as a memorial to the war dead of Leicestershire and Rutland, a University College had been opened in Leicester in the disused premises of the Leicestershire and Rutland County Lunatic Asylum. In 1925, on the strength of a good pass in the Intermediate BSc examination two years previously, the College (whose syllabuses and examinations were laid down by the University of London until it gained independent university status in 1957) awarded Snow a scholarship to complete an honours BSc in chemistry. In fact, the College had only appointed its first physical science lecturers in March of the same year, when Louis Hunter, an organic chemist and hydrogen bond expert, and A. C. (Sandy) Menzies, a spectroscopist, were appointed to lectureships in chemistry and physics respectively. Under their supervision, Snow studied pure mathematics, physics and chemistry as he was keen to pursue research. Although Hunter would later reminisce unkindly that Snow was the worst student of practical chemistry he had ever taught, Snow duly obtained a first class chemistry degree in 1927— the first student of the College to obtain first class in any discipline (Figure 42.1).

It was a requirement that candidates had to pass the practical examination but, according to the Leicester laboratory technician who moderated the examination, Snow failed the test but did such a brilliant theory paper that he was awarded first class honours. Departmental oral history has it that when the Leicester laboratory had been redecorated and proudly shown to visitors by Hunter, a few days later Snow performed a Skraup reaction. The contents of the flask were ejected through the reflux condenser and ended up on the newly painted ceiling. Hunter was furious and did not appreciate another student's suggestion that Snow be made to

Figure 42.1   Charles Percy Snow (1905–1980) playing the role of a blind wise man in a University College Leicester student production of Lady Gregory's Irish drama, *The Dragon* (1920), in 1926. (University of Leicester Archives)

climb up and autograph his handiwork. Hunter's solution was the edict that an asbestos board always had to be clamped above a reflux condenser! Snow was otherwise a typical student. He took part in college plays, played cricket, drank lots of beer and once wore a top hat for a month to win a bet (Figure 42.2).

Menzies, one of the country's first physicists to take up Raman spectroscopy, now urged Snow to specialize in spectroscopy, which was one of the twelve subjects the University of London allowed external students to study for the MSc by written and practical examination. As Menzies's very first research student, Snow investigated "the absorption of light by molecular films of cinnamic acid, and absorption spectra and chemical constitution", and duly obtained his MSc in 1928. This work was not awarded a distinction, but because Snow was the only external MSc student across the entire country for that year, the London University examiners generously awarded him a scholarship worth £200 per annum. This studentship was implicitly awarded for research at the University of London. However, Menzies, a former Fellow of

Figure 42.2   University College Leicester staff and students in 1927. A gowned C. P. Snow is first left in the seated row. Louis Hunter, his chemistry mentor, is fourth from the left in the same row. Photograph by A. Laxton Hames. (University of Leicester Archives)

Christ's College, Cambridge, was able to persuade the awarders to allow Snow to extend his spectroscopic research at the Department of Physical Chemistry at Cambridge; furthermore, Menzies arranged for Snow to be admitted to Christ's College. Clearly Snow's admission to the threshold of the "corridors of power" (an expression he made popular in his later novels) owed as much to Menzies as to his own considerable intellect.

WHY SCIENCE?

At this point it is worth pausing to ask why Snow had embarked upon a scientific career? In later life, he always insisted that "by vocation I was a writer". In a frank and revealing series of interviews he gave in 1978 to the American professor of English literature, John Halperin, Snow was explicit that as a teenager he had become convinced that one day he would be a novelist. There was never a burning desire to become a scientist; rather, it seemed the easiest way for him to make a living in the 1920s. Science came easily to him. "I should never have made a *good* scientist", he told Halperin, "but I should have made a perfectly adequate one".

Snow was not being merely arrogant in these interviews, for we know that he had written a novel, variously referred to as *The Devoted* and *Youth Searching,* in his first year at the University College of Leicester in 1925 or 1926, but destroyed it. So, we must picture him leaving Leicester in 1928 and proceeding to Cambridge to read for a PhD degree with no intention of pursuing a scientific career except insofar as it would give him financial support for his writing.

As it turned out, Snow was to publish some twenty-two research papers, as well as two mathematical chapters on molecular mass spectroscopy for the revised edition of Francis Aston's important textbook on isotopes. Of the twenty-two papers, sixteen were collaborations (six in status of research student, four as a research supervisor, and six as genuine collaborations with the Tasmanian physical chemist, Frank Philip Bowden), giving him a total of twelve original papers (54 per cent).

CAMBRIDGE

When Snow arrived in Cambridge, chemistry still formed part of what contemporaries called "The Papal State" because the titular and administrative head of the department was the stereochemist, Sir William Jackson Pope. Appointed in 1909 to develop a chemistry research school which would dispel Cambridge's long tradition of liberal, rather than specialized, teaching, Pope had long since abandoned research when Snow arrived. Endowments from British oil companies and from the Rockefeller Foundation immediately after the First World War had enabled Pope, and his chief lieutenant W. H. Mills, to expand the laboratories and to appoint Thomas Martin Lowry to a chair of physical chemistry. It was Lowry, rather than Pope, who administered the department, though all decisions of consequence remained Pope's. As Snow remarked later, Lowry had "with a curious kind of obstinacy, got stuck with researches on optical rotation that didn't attract many pupils".

On the other hand, research students (including Bowden and Snow) were attracted like moths to Eric K. Rideal, who had been appointed to the newly endowed Humphry Owen Jones Lectureship in Physical Chemistry in 1920. As Snow noted later, Rideal (the model for "Desmond" in *The Search*) "was willing to accommodate research on any topic from pure physics to biology, and his sub-department accordingly became a kind of hold-all for anyone who thought he had a decent problem". In practice, Rideal's programme was never as ramshackle as this may suggest and, at that time, Snow was well aware that Rideal's aim in encouraging diverse research in infra-red spectroscopy, electrochemistry, photochemistry, colloids and surface chemistry was his concern to understand the ultimate nature of chemical reactivity.

SNOW'S PERCEPTION OF PHYSICAL CHEMISTRY

In an undeservedly obscure essay on the nature of chemical research being conducted at Cambridge, which he published in 1933, Snow (by then Dr Snow, Fellow of Christ's College, Cambridge) perceptively analysed the direction in which chemistry was moving. The essay represents Snow's first, and very successful, attempt to explain the nature of scientific activity in lay terms—an accomplishment which was to become important to him as editor of the popular illustrated science monthly, *Discovery*, between 1938 and its wartime closure in March 1940.

"Chemistry", Snow defined, "is essentially the science of molecules". Atoms combined to form molecules, he explained, through the exchange of their outer electrons. Consequently, all modern chemists (and not merely those at Cambridge) were concerned with two theoretical problems: the structure of these molecules and their reactions. The primary tool for unravelling the structure of molecules was spectroscopy, which was likened by Snow to "solving an immense and rather tedious crossword puzzle", the process being "laborious in the extreme [albeit] exciting and invaluable in results". Although most work on simple diatomic molecules was being done in America (Snow was thinking primarily of the brilliant work of Robert S. Mulliken), the Cambridge workers were optimistic that "a physical invasion" of the electronic states of complicated organic molecules might soon be possible.

Cambridge chemistry had a long tradition of investigating the nature of complicated molecules, whether through Pope's and Mills's stereochemical researches, or Lowry's studies of the optical phenomena associated with molecular asymmetry (the subject in 1930 of Dorothy L. Sayers's detective story, *The Documents in the Case*). By the 1930s, however, Rideal's group had moved away from traditional chemical approaches in which the essential procedure was to take a molecule whose structure was known and then methodically alter it step-by-step into the molecule that was being investigated. Instead of a number of experiments being performed, followed by a single logical argument, Rideal's physical method usually employed one simple experiment followed by a theoretical argument. Snow argued strongly that the future of chemistry rested with physics. On that basis he was optimistic that "a physical invasion" of the electronic states of complicated organic molecules would soon be possible. As an example of the new methodology Snow instanced Rideal and Schulman's elegant experimental routine of compressing a unimolecular surface in order to derive the values of surface electric moments (potentials) of long-chain fatty acids. Snow shared Rideal's view that the Cambridge work on surfaces might eventually be a key to understanding the mechanism of chemical reactions.

Snow, who was never the most modest of men, intriguingly did not mention his own work in this 1933 review. He had published four papers in 1928 on the infra-red investigation of molecular structure. This work had included building a new piece of equipment with A. M. Taylor (then a fellow graduate student of Rideal's) which was a refinement of that used by the government chemist [Sir] Robert Robertson—notably employing a diffraction grating to obtain greater dispersion. The work had to be done at night after the university's electric power station had closed. Using this apparatus, he and Rideal investigated the diatomic molecules, nitric acid and carbon monoxide, and the relationship between Raman lines and infra-red bands.

Besides the advancement in infra-red apparatus, which was not outclassed until the development of instrumental technology by Randall, Barker and Meyer at the University of Michigan in the 1930s, Snow's originality in this work was to reveal a significant electronic band component in infra-red adsorption besides those hitherto recognized as due to oscillation, rotation and combined vibration–rotation. It was this finding which led him to participate in and to present a paper at an important Faraday Society discussion of "molecular spectra in relation to molecular structure" which Martin Lowry organized at the University of Bristol in September 1929. This meeting attracted spectroscopists from all over the world, including Raman, Hund, Herzberg, Taylor and Mulliken (who struck up a correspondence with Snow).

By 1929 spectroscopic nomenclature had been clarified by Mulliken, who had also developed the transition rules and affixed the quantum numbers needed to describe the electronic, vibrational and rotational states of diatomic molecules. All this Snow had imbibed and exploited, and from there he had gone on to see that, in principal, the same quantum mechanical techniques might be used to analyse the more complicated spectra of polyatomic molecules. This was where Snow was to come unstuck, for although he and others soon found that the study of molecules any more complicated than ethane, formaldehyde and dinitrogen tetroxide "insuperably difficult", he and his lifetime Australian friend, Philip Bowden, decided in 1930 that it ought to be possible in principle to crack the structures of biological molecules using infrared spectroscopy. (Bowden was to become the prototype for Frank Getliffe, the wise and sensitive scientist who appears in the *Strangers and Brothers* sequence of novels.)

In his essay on Cambridge chemistry, Snow wrote as follows on this biological work which used molecular spectroscopy and photochemical techniques to look at molecules of biological significance—notably, the vitamins. (Wherever, Snow wrote of Bowden only, I have deliberately pluralized relevant verbs and substituted the pronoun "we" in order to emphasize the collaborative nature of this work.)

"Usually, one particular group of two or more atoms, bound in a special way by their outer electrons, is essential for the kind of biological activity which is possessed by the vitamins, hormones and similar molecules. By using specially selected [monochromatic] light-rays, we alter these special groups in ways which we can detect by the aid of molecular spectroscopy. Even for large and complex molecules, the characteristic kind of light which they can give out is a valuable clue to the nature and behaviour of the active group. One interesting improvement used by us, which may very soon become a stock chemical method, is to freeze the molecules to a very low temperature. At low temperatures the molecules do not move about so freely; the characteristic kinds of light which they give out depend upon motions of the molecules, and become simpler the less the molecules are allowed to move."

It is not difficult to understand why Snow and Bowden moved into biochemistry. Rideal's fundamental work on surfaces had been inspired by William Bate Hardy's physiological investigations of colloids and membranes; the Gowland Hopkins's biochemical school at the Cambridge Dunn Institute was engaged in work on vitamins and nutrition—a field which was also the subject of spectroscopic study at Liverpool by Heilbron and Morton. At Liverpool, too, the spectroscopist E. C. C. Baly was making spectacular claims (later judged spurious) that he had replicated photosynthesis using ultra-violet radiation. Finally, at Wisconsin, Henry Steenbock had associated vitamin A activity with carotene and shown that rats fed upon a rickets-inducing diet benefited from exposure to sunlight or ultra-violet radiation. More surprisingly, Steenbock found that if rickets-producing diets containing a fraction of fat were radiated, rickets was prevented. The University of Wisconsin patented this discovery in 1925 and licensed food manufacturers to enrich foods with vitamin D by this method, using the dividends to pay for further nutritional research. Clearly there was money to be made from photochemistry, though we must also note the newly discovered Raman effect had made the irradiation of substances interesting.

THE PHOTOCHEMISTRY OF VITAMINS

On 14 May 1932 *Nature* readers were startled by a lead letter from Bowden and Snow entitled "The Photochemistry of Vitamins A, B, C and D". The two co-workers had identified the absorption spectra of vitamins and then irradiated a vitamin sample with light of a similar wavelength to that of one of the absorption bands. (Bowden developed a quartz monochromatic for this purpose in 1931.) Irradiation destroyed the electronic system producing the band and exchanged it for a new one, its special biological activity being destroyed. Hence, the two men argued, if suitable pre-vitamins could be irradiated with monochromatic light so that the correct vitamin bands were produced, the way would be open to synthesize vitamins by photochemistry, while also allowing some of a vitamin's structure to be determined. When β-carotene extracted from carrots was examined, Bowden and Snow identified a band at 2,700 Å [270 mμ]. They irradiated the sample with the mercury line 2650, and after some hours they identified a new band that had formed in the carotene at 3280 which was characteristic of vitamin A. Indeed, the solution responded to the antimony chloride test for the vitamin by producing a blue coloration. Animal feeding tests, they announced, would be needed before they could be absolutely confident that they had found a photochemical method for synthesizing the vitamin from a vegetable source. Although letters to *Nature* were not always refereed in the 1930s, that this was a piece of peer-reviewed approved research is indicated by the authors' acknowledgements to F. G. Hopkins, J. B. S. Haldane, N. W. Pirie, L. Harris and others.

The publication was a thirty days' wonder, receiving much publicity in *Industrial and Engineering Chemistry* ("production of vitamin A on a large scale and its manufacture in foods, such as bread and cereals, may be expected if recent British experiments are confirmed"), *Chemical News* ("by arranging for the exclusion of radiations which destroy the vitamin and free passage for those which create it, further yields should be obtained"), *The Lancet* and the *British Medical Journal* ("they believe they have produced vitamin A artificially"), as well as in the daily press. Yet, on turning to the Chemical Society's *Annual Reports* or to the *Reports of Progress in Applied Chemistry* for 1932, this sensational matter goes unreported. What had gone wrong?

As we now know, the conversion of β-carotene (provitamin A) into vitamin A depends upon enzyme action; but this was not the principal reason why Bowden and Snow's work was erroneous. One month after their announcement, on 11 June 1932, Ian Heilbron of Liverpool, together with his student, Richard Morton, published a devastating refutation of the Cambridge work in *Nature*. They made five points.

1. The Cambridge pair's technique was less original than Bowden and Snow had claimed. Indeed, Morton himself had used spectral analysis to identify vitamin A_2 in 1928.
2. Their assumption that *one* experimental technique was the key to the structure of several different vitamins was to underestimate the differences between organic compounds of widely different constitutions and very different physiological actions.
3. Although β-carotene was a metabolic precursor of vitamin A, it did not follow that it was a photochemical precursor.
4. Bowden and Snow's irradiated carotene still contained carotene, "so that the relevance of the promised biological assay does not emerge since it has been established that carotene is converted *in vivo* into vitamin A".
5. The authors had ignored a more obvious explanation for their observations, namely molecular rearrangement, and their speculations were either "so true as to be obvious" or "uninformative and premature".

Despite further criticism from research groups at Cincinnati and Cambridge itself, Bowden and Snow never admitted publicly that their reasoning had been erroneous, although they did admit a year later, in a paper on Bowden's low temperature absorption spectroscopy, that the peaks of irradiated carotene and vitamin A were not the same. Nothing more was said about the synthesis of vitamins here or in the definitive account of their experiments read to the Royal Society on 26 April 1934. Heilbron and Morton's criticism was acknowledged and readers were left to draw their own conclusions.

THE NOVELIST

If this embarrassing and premature announcement was not sufficient
to turn Snow away from scientific research, further work in 1934
while supervising Eric Eastwood must have been the last straw. In
May 1934 he and Eastwood argued in *Nature* that similar ultra-
violet absorption bands occurred in homologous series of aldehydes.
Within six months, however, they had to admit that the similarities
were due to the presence of a benzene impurity carried over from the
catalyst used in the preparation of the aldehydes. Moreover, the
quartz plates of their apparatus had caused interference.

Had Snow left too much responsibility to Eastwood because his
time was absorbed in writing novels? Significantly, although the
work was written up for the Royal Society's *Proceedings* in 1935,
Snow did no more research after March of that year when he was
made a Tutor at Christ's College—a teaching post he kept formally
until 1945. (One cannot help suspecting that the paper was read to
the Royal Society, rather than the better suited Chemical Society
because Snow had ambitions to become a Fellow of the Royal
Society. He was never proposed.)

By then Snow had already published three novels—the anonymous detective story *Death under Sail* (1932, reprinted 1966), the
anonymous utopian *New Lives for Old* (1933), and *The Search*
(1934, revised 1958) which, as Snow admitted in later life, capita-
lized on the abortive vitamin work. In this excellent study of scientific careerism, the hero Arthur Miles who works at the frontiers
of crystallography (not spectroscopy) blunders because he accepts
information from a collaborator without checking. He also breaks
the scientific moral code as a sign that he has finished with science
by not exposing a friend's fraudulent work. Joseph Needham,
reviewing *The Search* in *Nature*, called it "a crystallographic
Martin Arrowsmith" (the reference was to Sinclair Lewis's
acclaimed bacteriological novel of 1922) and encouraged Snow to
further authorship with the words "the results of such a life work
are ... no less valuable than 100 pages in *Proceedings* and *Transactions*". Six years' later Snow ensured his fame would rest in literature rather than in science by the publication of the first book in
the nonogenal sequence of novels, *Strangers and Brothers*.

Despite the decision to make his name as a writer, Snow always
remained close to his scientific roots. During the Second World

War he became a civil servant placed in charge of directing scientific and technical personnel for the Ministry of Labour, a task he continued in peace time both for the English Electric Company and for the Civil Service Commissioners. Knighted in 1957, Harold Wilson's Labour government made him a life peer in 1964 when he was appointed Parliamentary Secretary for the short-lived Ministry of Technology. All this time he was writing and publishing the Trollopian novels which follow the career of his *alter ego*, Lewis Eliot, through the trials and tribulations of the twentieth century. He died on 1 July 1980 remembered, as he would have wished, as a novelist rather than as a spectroscopist.

Further Reading

Publication details are given only for works not published in either the UK or USA.

NB All web links were accessed in early May 2011.

PREFACE

C. A. Russell, 'Rude and disgraceful beginnings: a view of history of chemistry from the nineteenth century', *British Journal for the History of Science*, 1988, **21**, 273–294.

David Knight, *Ideas in Chemistry: A History of the Science*, 1992.

William H. Brock, *The Fontana History of Chemistry*, 1992 (published in the USA as *Norton History of Chemistry*, 1993, and as *The Chemical Tree. A History of Chemistry*, 2000).

David Knight and Helge Kragh, ed., *The Making of the Chemist. The Social History of Chemistry in Europe, 1789–1914*, 1998.

Peter Morris, 'Writing the history of modern chemistry', *Bulletin for the History of Chemistry*, 2007, **32**, 2–9.

CHAPTER 1

The Times, 30 September 1868, pp. 3, 5–8 and 16 October 1868.

Maurice R. Fox, *Dye-Makers of Great Britain 1856–1976*, 1987.

Jeremy Farrell, *Socks & Stockings*, 1992.

W. H. Brock, *William Crookes (1832–1919) and the Commercialization of Science*, 2008.

The Case of the Poisonous Socks: Tales from Chemistry
By William H. Brock
© William H. Brock, 2011
Published by the Royal Society of Chemistry, www.rsc.org

CHAPTER 2

W. H. Brock, *Protyle to Proton. William Prout and the Nature of Matter 1785–1985*, 1985.

Yoshiyuki Kikuchi, 'Redefining academic chemistry: Jōji Sakurai and the introduction of physical chemistry into Meiji Japan', *Historia Scientiarum*, 2000, **9**, 215–256.

Jonah Lehrer, *Proust was a Neurosurgeon*, 2007.

Amy Leyva, 'A taste sensation', *Chemical Heritage*, 2008, **26**, 37.

CHAPTER 3

There are several essays in English in Christoph Meinel and Hartmut Scholz, ed., *Das Allianz von Wissenschaft und Industrie. August Wilhelm Hofmann (1818–1892), Zeit, Werk, Wirkung*, VCH, Weinheim, 1992, including my own 'Liebig and Hofmann's impact on British culture'.

CHAPTER 4

W. H. Brock, *Justus von Liebig. The Chemical Gatekeeper*, 1997 (paperback, 2002).

Harold McGee, *On Food and Cooking. The Science and Lore of the Kitchen*, 1984 (paperback 1984).

CHAPTER 5

Adapted from K. Bayertz and R. Porter, ed., *From Physico-Theology to Bio-technology: Essays in the Social and Cultural History of Biosciences*, Rodopi, Amsterdam, 1998, pp. 88–107. Quotations from the writings of Hodgkins are by permission of Smithsonian Institution Archives.

Kenneth Hafertepe, *America's Castle. The Evolution of the Smithsonian Building and its Institution, 1840–1878*, 1984.

Gwen Caroe, *The Royal Institution. An Informal History*, 1985.

Frank A. J. L. James, ed., *'The Common Purposes of Life'. Science and Society at the Royal Institution of Great Britain*, 2002.

CHAPTER 6

Based on a talk given to the Society for the History of Alchemy and Chemistry in 2000.

Mark S. Morrisson, *Modern Alchemy. Occultism and the Emergence of Atomic Theory*, 2007.

For occult chemistry, including illustrations, see www.chem.yale.edu/~chem125/125/history99/8Occult/OccultAtoms.html.

CHAPTER 7

Adapted from a talk given to the Society for the History of Alchemy and Chemistry on 10 December 2007.

R. A. Gilbert, *A. E. Waite. Magician of Many Parts*, 1987.

'Exploring alchemy in the early 20th century', online article compiled by Cis van Heertum in February 2006 for the J. R. Ritman Library, Bibliotheca Philosophica Hermetica, Amsterdam, and available at www.ritmanlibrary.nl/c/p/h/bel_18.html.

George B. Kauffman, ed., *Frederick Soddy (1877–1956)*, 1986.

Alex Owen, *The Place of Enchantment. British Occultism and the Culture of the Modern*, 2004.

Mark S. Morrison, *Modern Alchemy. Occultism and the Emergence of Atomic Theory*, 2007.

CHAPTER 8

From a talk given to a joint meeting of the Prince Albert Society and the Victorian Society at the Victoria & Albert Museum in July 1999 and published in Franz Bosbach, ed., *Prince Albert and the Development of Education in England and Germany in the 19th Century*, K. G. Saur, Munich, 2000.

H. Spencer, *Education Intellectual, Moral and Physical*, 1854 (with many reprints).

David Layton, *Science for the People*, 1973.

E. W. Jenkins, *From Armstrong to Nuffield. Studies in Twentieth-Century Science Education in England and Wales*, 1979.

CHAPTER 9

Adapted from *Ambix*, 1967, **14**, 133–139.

T. S. More and J. C. Philip, *The Chemical Society 1841–1941*, 1947.

C. A. Russell, N. G. Coley and G. K. Roberts, *Chemists by Profession. The Origins and Rise of the Royal Institute of Chemistry*, 1977.

Gwen Averly, 'The 'social chemists': English chemical societies in the eighteenth and early nineteenth centuries', *Ambix*, 1986, **33**, 99–128.

Trevor Levere and Gerard L. E. Turner, *Discussing Chemistry and Steam. The Minutes of a Coffee House Philosophical Society 1780–1787*, 2002.

C. A. Russell, 'The Marreco story' in *Chemistry, Technology and Society*, Proceedings of the Fifth International Conference on the History of Chemistry, held Estoril and Lisbon, 6–10 September 2005, Aveiro, Portugal, 2006, pp. 374–383.

CHAPTER 10

Adapted from *Ambix*, 2000, **47**, 121–134.

Gerda Elizabeth Bell, *Ernest Dieffenbach, Rebel and Humanist*, Dunmore Press, Palmerston North, New Zealand, 1976.

Gerrylynn K. Roberts, 'The establishment of the Royal College of Chemistry: an investigation of the social context of early Victorian chemistry', *Historical Studies in the Physical Sciences*, 1976, **7**, 437–485.

CHAPTER 11

Originally published as 'Les nouvelles cathedrals de la science' in *Les Cahiers de Science & Vie*, no. 51, June 1999.

W. H. Brock, *Justus von Liebig. The Chemical Gatekeeper*, 1997 (paperback, 2002).

Frank A. J. L. James, ed., *The Development of the Laboratory. Essays on the Place of Experiment in Industrial Civilisation*, 1989.

CHAPTER 12

Based on the Dexter Award lecture given at the meeting of the American Chemical Society, Chicago, 22 August 1995, and published in *The Bulletin for the History of Chemistry*, 1998, **21**, 1–11.

R. Sviedrys, 'The rise of physical laboratories in Britain', *Historical Studies in the Physical Sciences*, 1976, **7**, 405–436.

Graeme Gooday, 'Precision measurement and the genesis of physics teaching laboratories in Victorian Britain', *British Journal for the History of Science*, 1990, **23**, 25–51.

CHAPTER 13

M. P. Crosland, *Historical Studies in the Language of Chemistry*, 1962 (reprint Dover paperback, 1978).

W. H. Brock, 'The British Association Committee on chemical symbols 1834', *Ambix*, 1986, **33**, 33–42.

J. L. Alborn, 'Negotiating notation: chemical symbols and British society, 1831–1835', *Annals of Science*, 1989, **46**, 437–460.

Ursula Klein, *Experiments, Models, Paper Tools: Cultures of Organic Chemistry in the Nineteenth Century*, 2003.

CHAPTER 14

Alexander Scott (Russell's son-in-law who inherited the B-Club's papers and deposited them in the RSC library), Presidential Address, *Journal of the Chemical Society*, 1916, 342–351.

J. G. Paradis, 'Satire and science in Victorian culture' in B. Lightman, ed., *Victorian Science in Context*, 1997, pp. 143–175.

Mara Beller, 'Jocular commemorations: the Copenhagen spirit', *Osiris*, 1999, **14**, 252–273.

CHAPTER 15

For Gordon van Praagh, see *Oxford Dictionary of National Biography* (online addition 2008). For Chinese source, see http://en.wikiquote.org/wiki/Chinese_proverbs.

CHAPTER 16

Adapted from *Education in Chemistry*, 2006, **43**, 106–108.

N. G. Coley, 'The physico-chemical studies of Amedeo Avogadro', *Annals of Science*, 1964, **20**, 195–210.

Nicholas Fisher, 'Avogadro, the chemists, and historians of chemistry', *History of Science*, 1982, **20**, 77–102, 212–231.

Mario Morselli, *Amedeo Avogadro*, Reidel, Dordrecht, 1984.

Gayle Brickert-Albrecht and Dan Morton, 'A biographical interview with Lorenzo Romano Amedeo Carlo Avogadro', www.woodrow.org/teachers/ci/1992/Avogadro.html.

CHAPTER 17

From an unpublished lecture given at a conference on the development of science in Göttingen held at the Georg-August-Universität-Göttingen, 25 November, 2000.

J. L. W. Thudichum, 'On the discoveries and philosophy of Liebig, with especial reference to their influence upon the advancement of arts, manufactures and commerce', *Journal Society of Arts*, 1875–76, 80–86, 95–100, 111–116, 141–145.

O. T. Benfey, ed., *Classics in the Theory of Chemical Combination*, 1963.

Robin Keen, *The Life and Work of Friedrich Wöhler*, Edition Lewicki-Büttner, Verlag Traugott Bautz, Nordhausen, 2005.

Ursula Klein, *Experiments, Models, Paper Tools. Cultures of Organic Chemistry in the Nineteenth Century*, 2003.

CHAPTER 18

Adapted from *Endeavour*, 1996, **20**, 121–125.

C. A. Russell, *History of Valency*, 1971.

A. J. Rocke, 'Subatomic speculations and the origin of structure theory', *Ambix*, 1983, **30**, 1–18.

J. H. Wotiz, ed., *The Kekulé Riddle. A Challenge for Chemists and Psychologists*, 1993.

A. J. Rocke, *Image & Reality. Kekulé, Kopp, and the Scientific Imagination*, 2010.

CHAPTER 19

Adapted from *Hyle*, 2002, **8**, 49–54.

W. H. Brock, ed., *The Atomic Debates. Brodie and the Rejection of the Atomic Theory*, 1967.

CHAPTER 20

Adapted from *Chemistry in Britain*, 1996, **32**, 37–39.

J. Vargas Eyre, *Henry Edward Armstrong 1848–1937*, 1958.

W. H. Brock, *H. E. Armstrong and the Teaching of Science 1880–1930*, 1973.

W. H. Brock, 'The laboratories of Finsbury Technical College', in Frank A. J. L. James, ed., *The Development of the Laboratory*, 1989, pp. 155–170.

CHAPTER 21

Adapted from a book review in *Chemistry & Industry*, July 2004.
Tony Harrison, *Square Rounds*, 1992.
Dietrich Stoltzenberg, *Fritz Haber. Chemist, Nobel Laureate, German, Jew*, 2004.

CHAPTER 22

Adapted from *The Bulletin for the History of Chemistry*, 2009, **34**, 11–20.
J. R. Partington, 'Nernst Memorial Lecture', *Journal of the Chemical Society*, 1953, 2853–2872.
G. P. Moss and M. V. Saville, *From Palace to College. An Illustrated Account of Queen Mary College*, 1985.

CHAPTER 23

Anon, 'Death of Henry Crookes', *Nature*, 1915, **96**, 11–12, 47.
Anon, 'Henry Crookes', *Journal of the Institution of Electrical Engineers*, 1916, **54**, 677.
Henry Crookes, 'On metallic colloids and their bactericidal properties. The history of collosols', *Chemical News*, 1914, **109**, 217–219.
Alfred Broadhead Searle, *The Use of Colloids in Health and Disease*, 1920, which is illustrated with photographs of Henry Crookes's 'before and after' cultures.
For the High Court action see *Chemist & Druggist*, July–December 1914, **85**, 81, 216, 342.
For the dispute over colloidal state see *British Medical Journal*, 1917, **1**, 617 and *British Medical Journal*, 1923, **1**, 273.
Howard Greenfeld, *The Devil and Dr Barnes. Portrait of an American Art Collector*, 1987.
W. H. Brock, *William Crookes (1832–1919) and the Commercialization of Science*, 2008.

CHAPTER 24

Adapted from a book review in *Chemistry & Industry*, January 2001.

George A. Olah, *A Life of Magic Chemistry: Autobiographical Reflections of a Nobel Prize Winner*, 2001.

CHAPTER 25

Raphael Patei, *The Jewish Alchemists*, 1994.

Tara Nummedal, 'Alchemical reproduction and the career of Anna Maria Zieglerin', *Ambix*, 2001, **48**, 56–68.

Tara Nummedal, *Alchemy and Authority in the Holy Roman Empire*, 2007.

CHAPTER 26

Margaret Hill and Alan Dronsfield, 'Ida Freund, the first woman chemistry lecturer', *Education in Chemistry*, September 2004. See also Freund's entry in *Oxford Dictionary of National Biography*, 2004.

Pnina Abir-Am and Dorothy Outram, ed., *Uneasy Careers and Intimate Lives: Women in Science 1789–1979*, 1987 (focuses on married women scientists).

Mary R. S. Creese, 'British women of the nineteenth and early twentieth centuries who contributed to research in the chemical sciences', *British Journal for History of Science*, 1991, **24**, 275–305.

Joan Mason, 'The forty years war to admit women', *Chemistry in Britain*, 1991, **27**, 233–238.

Geoffrey and Marelene Ranham-Canham, *Chemistry was their Life. Pioneer British Women Chemists, 1880–1949*, 2008.

Melinda Baldwin, 'Where are your intelligent mothers to come from?' Marriage and family in the scientific career of Kathleen Lonsdale FRS (1903–71)', *Notes & Records of the Royal Society*, 2009, **63**, 81–94.

CHAPTER 27

Adapted from a book review in *Chemistry & Industry*, October 2002.

Eugene G. Rochow and Eduard Krahé, *The Holland Sisters: Their Influence on the Success of their Husbands Perkin, Kipping and Lapworth*, 2001.

CHAPTER 28

From an unpublished lecture given at the Sir Christopher Ingold Symposium at the University of Southampton on 28 October 1993.

Kenneth T. Leffek, *Sir Christopher Ingold. A Major Prophet of Organic Chemistry*, 1996, especially chapter 5.

Marelene and Geoffrey Rayner-Canham, 'A tale of two spouses' [on Gertrude Walsh and Hilda Usherwood], *Chemistry in Britain*, 1999, **35**, 45–46.

CHAPTER 29

Thomas Hapke, *Die Zeitschrift für physikalische Chemie. Hundert Jahre Wechselwirkung zwischen Fachwissenschaft, Kommunikationsmedium und Gesellschaft*, Traugott Bautz, Hertzberg, 1990.

Wolfgang Caesar, 'Liebigs Annalen. Ein Nachruf', *Deutsche Apotheker Zeitung*, 1998, **138**, pp. 52–59.

Wilhelm Fresenius, '140 years 'Fresenius' Journal of Analytical Chemistry', *Fresenius Journal of Analytical Chemistry*, 2001, **371**, 1041–1042.

CHAPTER 30

Adapted from *Paradigm. Journal of the Textbook Colloquium*, 2001, **2**, 2–5 [publication has since ceased].

W. H. Brock and A. J. Meadows, *The Lamp of Learning. Taylor & Francis and the Development of Science Publishing*, 1984, 2nd edn, 1998.

Anders Lundgren and Bernadette Bensaude-Vincent, ed., *Communicating Chemistry: Textbooks and their Audiences, 1789–1939*, 2000.

CHAPTER 31

See entry on Watts in *Oxford Dictionary of National Biography*, 2004.

W. H. Brock, 'The society for the perpetuation of Gmelin', *Annals of Science*, 1978, **35**, 599–617.

CHAPTER 32

The Gosse and Thompson letters form part of the Warington archive at Rothamsted Research, Hertfordshire, and are quoted by kind permission of the Lawes Agricultural Trust.

Adapted from *Archives of Natural History*, 1991, **18**, 179–183.

David E. Allen, *The Naturalist in Britain. A Social History*, 1976.

Nick Baker, 'William Thompson – the world's first underwater photographer', *Historical Diving Times*, Summer 1997, Issue 19, available at www.thehds.com/publications/thompson.html.

Ann Thwaite, *Glimpses of the Wonderful. The Life of Philip Henry Gosse*, 2002.

CHAPTER 33

E. O. Sachs, *Modern Opera Houses and Theatres*, three volumes, 1896–1998; reprinted New York, 1968.

Edwin O. Sachs Photographic Collection, Royal Opera House, Covent Garden, London.

G. V. Blackstone, *A History of the British Fire Service*, 1957.

Victor Bailey, ed., *Forged in Fire: The History of the Fire Brigades Union*, 1992.

David J. Collins, 'An explosion of nitroglycerine in Sydney: A window on chemistry in mid-nineteenth century Australia', *Historical Records of Australian Science*, 2002, **13**, 131–149.

CHAPTER 34

For another ingenious mathematical chemist see James Walker, 'Alexander Crum Brown', *Journal of the Chemical Society*, 1923, **123**, 3422–3431.

For a useful historical overview see Nenal Trinajstic and Ivan Gutman, 'Mathematical chemistry', *Croatica Chemica Acta*, 2002, **75**, 329–356, available at www.vnovak.hr/ccacaa/CCA-PDF/cca2002/v75-n2/CCA_75_2002_329_356_TRINAJS.pdf.

CHAPTER 35

J. W. Mellor, 'Some chemical and physical changes in the firing of pottery', *Journal of Society of Chemical Industry*, 1907, **26**, 375–379.

J. W. Mellor, *Uncle Joe's Nonsense*, 1934.

Fathi Habashi, 'Joseph William Mellor (1869–1038)', *Bulletin for the History of Chemistry*, 1990, **7**, 13–16.

CHAPTER 36

A useful sketch of the Baker lecturers is given in the reminiscences of A. W. Laubengayer published in: *Department of Chemistry Newsletter*, March 1976, Issue 18, 5–10, available at http://ecommons.cornell.edu/bitstream/1813/3087/1/CCB_018.pdf; and *Department of Chemistry Newsletter*, August 1976, Issue 19, 5–14, available at http://ecommons.cornell.edu/bitstream/1813/3088/1/CCB_019.pdf.

E. Heuser and B. W. Rowland, 'Dr Johan Rudolf Katz, 1880–1938', *Journal of Chemical Education*, 1939, **16**, 153–154.

CHAPTER 37

Adapted from *Chem@Cam*, Spring 2004, p. 16.

John Shorter, 'Chemistry at Cambridge under George Liveing' in Mary D. Archer and Christopher D. Haley, ed., *The 1702 Chair of Chemistry at Cambridge*, 2005, chapter 7.

W. H. Brock, 'Thomas Lamb Phipson' in B. Lightman, ed., *Dictionary of Nineteenth-Century British Scientists*, 2004, vol. 3, pp. 1597–1598.

CHAPTER 38

'Address to Students', *Chemical News*, 12 September 1879, **40**, 111. Huxley's address, one of dozens he must have given at school speech days, was not printed in his collected works.

The Dr Williams Library, 14 Gordon Square, London WC1H 0AR is home to the New College papers.

Ebenezer Prout lacks a definitive biography, but there is a fine entry in *New Grove Dictionary of Music and Musicians*, 1980.

CHAPTER 39

Chemical News, 27 October 1865, **12**, 206.
Chemical News, 16 November 1866, **14**, 238 and 286.
Chapman's obituary, *Journal of the Chemical Society*, 1872, **26**, 775.
Christopher Hamlin, *A Science of Impurity. Water Analysis in Nineteenth-Century Britain*, 1990.
Alan J. Rocke, *The Quiet Revolution. Hermann Kolbe and the Science of Organic Chemistry*, 1993.

CHAPTER 40

G. Du Maurier, 'Recollections of an English goldmine', *Once a Week*, 21 September 1861, **1**, 356–364.
Leonée Ormond, *George Du Maurier*, 1969, quotation from p. 24 by permission of Routledge & Kegan Paul [now Taylor & Francis], and her Du Maurier entry in *Oxford Dictionary of National Biography*, 2004.
David Lodge, *Author, Author*, 2004.

CHAPTER 41

H. S. Patterson, R. S. Cripps and R. Whytlaw-Gray, 'The critical constants and orthobaric densities of xenon', *Proceedings of the Royal Society*, 1912, **86A**, 579–590.
Sir George Schuster, 'Cripps', *Biographies of Members and Fellows of the Royal Society*, 1955, **1**, 11–32.
Simon Burgess, *Stafford Cripps: A Political Life*, 1999.
Peter Clarke, *The Cripps Version. The Life of Sir Stafford Cripps 1889–1952*, 2002 (paperback, 2003).

CHAPTER 42

Adapted from *Chemistry in Britain*, April 1988, 345–347 and 364.
C. P. Snow, 'Chemistry', in Harold Wright, ed., *Cambridge University Studies 1933*, 1933, pp. 97–127 (quotation from p. 108 by permission of the Snow literary estate administered by Curtis Brown).
C. P. Snow, The Reade Lecture, *The Two Cultures*, 1959.

F. R. Leavis, The Richmond Lecture, *Two Cultures? The Significance of C. P. Snow*, 1962,

C. P. Snow, *The Two Cultures: and a Second Look*, 1964.

For Hunter (1899–1986) see *Chemistry in Britain*, 1987, **23**, 878.

For Menzies (1897–1974), who became Professor of Physics at the University of Southampton in 1932, see *The Times*, 7 June 1974, p. 21.

Jerome Thayle, *C. P. Snow*, 1964, provides a good account of Snow's literary activities in the 1930s.

Neil Bezel, 'Autobiography and 'The Two Cultures' in the novels of C. P. Snow', *Annals of Science*, 1975, **32**, 555–571.

John Halperin, *C. P. Snow. An Oral Biography*, 1983; quotations by permission of Harvester Press [now Taylor & Francis].

For the background to Snow's spectroscopy, see R. Norman Jones, 'Analytical applications of vibrational spectroscopy – a historical review' in James R. Durig, ed., *Chemical, Biological and Industrial Applications of Infrared Spectroscopy*, 1985, pp. 1–50.

Subject Index

1846: state of chemistry in Britain 74–84
1901: future of chemistry 40–7

A Comprehensive Treatise on Inorganic and Theoretical Chemistry 267
A Naturalist's Ramble on the Devon Coast 253
A Second Visit to the United States of North America 77
A Short History of Everything 137
A Technological Dictionary of Fire Insurance 255–6
A Technological Fire Insurance Commentary 257
A Treatise on Quantitative Analysis…Clays, Silicates… 267, 269–70
Abdul-Ali, Sijil 54
Abel, Frederick 120
Abney, William de W. 65
Aborigines Protection Society 77
Accum, Frederick 20, 243
Acton, Eliza 21
Acton, Hannah 26, 30
Adams, W. G. 97

Advanced Physical Chemistry 185, 191
Advanced Treatise of Physical Chemistry 264
Agassiz, Louis 272
Agricultural Chemistry 143
Ajinomoto (*Aji-No-Moto*) company 15
Alchemical Society 1912–1915 48–54
Alchemistiche Blätter 54
Alchemy Ancient and Modern 53
"Alere flamman" 240
alka-hest (universal solvent) 202–3
Ambix 48
American Chemical Society 265
ammonia structure 287–8
Ampère, André-Marie 97, 134
An Address to the Agriculturalists of Great Britain… 79
An Introduction to Chemical Philosophy 195
Andrews, Thomas 97
Anfangsgründe der Stöchyometrie oder Messkunst chemische Elemente 262

Anglo-Saxon 243
"aniline violet" 6
Animal Chemistry 20, 116, 143
Animal Chemistry Club 67
Annalen der Chemie 79, 144, 233, 235
Annalen der Pharmacie 142
Annals and Magazine of Natural History 249–50
Annals of Natural History 254
Annals of Philosophy 73
Annual Reports (Chemical Society) 43, 230, 311
Anschütz, Richard 149, 156, 157
aquarium chemistry 248–54
Archimedes 124
Argyrol 199
Armstrong, Henry
  B-Club 120
  chemical critic 167–8
  "chemical engineering" 172
  Chemical News 168
  Chemical Society 169
  chemist in coloured waistcoat 174–5
  Chemistry & Industry 166
  chemistry teaching 211
  Dewar, James 35
  Frankland, Edward 168
  future of chemistry 46–7
  Huxley memorial lecture 113
  hydrogen sulfide 90
  Kekulé, August 168
  low temperature physics 26
  Perkin, William 216
  physical chemistry 42
  Pope, William 281
  research school 171–4
  Royal Institution 35–6, 174
  science education 124
  teaching of chemistry 43, 64–5, 103
  training 168–71
  Usherwood, Edith Hilda 229
  Usherwood, Thomas 219
  writer and critic 173
Arnett, Ned 203
Arrhenius 168, 178
arsenic (dye manufacture) 7
Association of British Chemical Manufacturers 260
Aston, Frances 3–5
Atkinson, Edmond 246
Atomic Structure of Minerals 274
Atwood, Mary Anne 50–1
Author, Author 289
Autobiography 102, 113
Avogadro, Amedeo 131–7
Avogadro's Number 131, 137
Ayrton, William 95, 101, 103, 110, 170–1

B-Club
  description 118–23
  Guthrie, Frederick and "on graphic formulae" 106
  members 122–3
Babbage, Charles 81
Bacillus phosphorescens 198
Baeyer, Adolf 151
Baker, G. F. 271
Baker, Henry 271
Baker, Herbert Brereton 173, 188
Baker Laboratory of Chemistry, Cornell 271–2
Baker lectures 271–5
Baly, E. C. C. 37, 309

Bancroft, W. D. 174
"Barefoot of Taunton" 6
Barger, George 273
Barnes, Albert C. 192
Barrett, William 111
Barth, Otto Wilhelm 54
Battersea Grammar School 54
Bayer 170
Beagle (ship) 77, 104
Beale, William Phipson 118
Bell, Jacob 81
Bence Jones, Henry 24–5
benzene 18, 155, 156, 174
"benzoyl hydride" 116
benzoyl radical 116
Berkshire Beauty Products 54
Bernal, J. D. 178
Berthelot, Daniel 296
Berthollet, Claude 135
Berzelius, Jöns 75, 80, 115–17,
    134, 139, 143–5
Besant, Annie 44–5
Besant, Walter 104–7
"better living with
    chemistry" 41
"bifurcation" (two or more
    science subjects) 63
Biological Chemistry
    Hoppe-Seyler 236
Birkbeck, George 67–8, 71–3
Birkbeck Laboratory, UCL 91
bitter almonds
    (benzaldehyde) 141
Black, Joseph 46
Blackwoods Magazine 29
Blavatsky, Madame 44, 49
blue dye 4
"blue stockings" 4
Bohemism 49
Boltwood, Bertram 46

Bonn Chemical Institute 157
Boole, George 162–3
Boots 201
Boranes in Organic
    Chemistry 274
Borrow, George 105
Bosch, Carl 41, 176
Bosworth, J. 243
Bovril 23
Bowden, Frank Philip 305–6,
    308–11
Boyle, Mary 221
Bragg, William Henry 189
Bragg, William Lawrence 274
Bramwell, Frederick 30
Brande, William 25, 73, 159
Bredig, Georg 14, 198
Brewster, David 80
British Association for the
    Advancement of Science
    (BAAS)
  Armstrong, Henry 171
  B-Club 118
  Berzelian alphabetical
    signs 116
  chemistry of the future 42
  hydrogen cyanide 225
  von Liebig, Justus 142
  organic nomenclature 99
  science education 63
  Young, Sydney 40–1
British Colloids Ltd 201
British Fire Prevention
    Committee 259
British Medical Journal 201, 310
British Science Guild 43
Brodie Jr, Benjamin Collins 80,
    158–65
Brodie Sr, Benjamin
    Collins 121, 147, 158

"Brodie's ozonizer" 164
Brønsted, Johannes 202
Brough, John Cargill 97, 118,
    120, 147
Brown, Herbert 203–4, 274
Brown, Horace 168
Brown, John Campbell 97
Browne, James Crichton 28–9
Brownian motion 198
Brunton, Thomas Lauder 104
Bryant, Sophie 220
Bryson, Bill 137
Buckland, William 78
Buff, Heinrich 151
Bullock, John Lloyd 78
Bunsen, Robert 93, 97, 99–100,
    151, 170, 247, 288
Burge, Hubert M. 295
Burgess, Simon 297
Butterfield, William 64

Cailletet, L. 297
Calculus of Chemical
    Operations 42, 162
Cameron, Alec Munro 259–60
camphene hydrochloride 203
Cannizzaro, Stanislao 136, 154
Cantor's Dilemma 214
carbocations 203
carbon tetravalence 152
carbylamine (HNC) 223–4
Carey Foster bridge 101
Caro, Heinrich 19
Carroll, M. F. 187
Carte, Richard 282
Cassell Cyanide Company 224
"Cathedrals of Science"
    (laboratories) 96
"cationotropy" 227
Cavendish Society 245

"central science" (chemistry) 56
CERAM research 269
Ceramics Society 268–70
Cerny, Frederick
    (pseudonym) 105
Chadwick, Edwin 79
Chambers, Dr M. L. 27, 32
Chapman, Ernest
    Threophon 288–9
"Charcoal Period" (Royal
    Institution) 36
Chartered Insurance
    Institute 258, 260
chemical algebra 114–17
Chemical Amusement 243
Chemical Catechism 242
Chemical Club 120
Chemical Essays 243
Chemical Gazette 241
Chemical Kinetics 274
Chemical Letters 21, 150
Chemical Manipulation 86
Chemical Method 151
Chemical News
    Armstrong, Henry 168
    Crookes, Henry 201
    Crookes, William 5, 73, 100,
        284, 286, 288
    Guthrie, Frederick 106
    Phipson, Thomas 279
    photochemistry of
        vitamins 310
    Redgrove, Herbert 48
    structural formulae 288
    William Rider & Co. 52
chemical origins of practical
    physics
    Foster, George Carey 99–102
    Guthrie, Frederick 102–13
    introduction 97–9

"Chemical Revolution" 85
Chemical Society
  Annual Reports 43,
    230, 311
  Armstrong, Henry 169
  Brodie Jr, Benjamin
    Collins 161
  Graham, Thomas 142
  London Chemical
    Society 67, 70
  Partington J. R. 187
  Ramsay, William 42
  science education 63
  soirées 122
  see also Royal College of
    Chemistry
*Chemical Statics and
  Dynamics* 189, 264, 268
chemical weapons 41
Chemische Briefe 143
Chemisches Journal 233
Chemistry & Industry 166, 168,
  175, 187
Chemistry Club 122
Chemistry in Commerce 15
Chemistry of Common Life 242
Chemistry of Discovery 125
chemistry by discovery in a
  phrase 124–5
*Chemistry of Fire and Fire
  Prevention* 258
"Chemistry of the Future" 42
*Chemistry in Relation to Fire
  Risk and Fire Extinction* 260
Chesney, George Tomkyns 29
Chevreul, Michel 281
chloroxynaphthalic acid
  (red dye) 6
Christ's Hospital school,
  Horsham 124, 171

Chronologie Chemie 1800–1980
  43
City of London School 241
Clark, James 78
Clarke, Ellen 289
Clarke, Frank
  Wigglesworth 112
Clerke, Agnes 35–6
Cohen, Ernst Julius 272
"collosol argentums" 200
collosols 198–9
*Comprehensive Treatise on
  Theoretical and Inorganic
  Chemistry* 190
Comte, Auguste 161
Conan Doyle, Arthur 214
Conant, James 203
*Concrete and Constructional
  Engineering* 259
Congreve, Richard 161
Consolations in Travel 158
Cook, Florence 193
Cooper, John Thomas 72–3
Cooper, K.E. 227
Corah, Nathaniel 4
Corke, Harriet 240
Cornell University, USA 271–3
Cosmos 279
Cotta, Johann Georg 77
Crell, Lorenz 233
Crell's Journal 233
Crimean War 65
Cripps, Stafford 294–300
"Cripps test" 299
Cronin, A. J. 214
Crookes, Alice 195
Crookes Colloids 201
"Crookes Collosols" 199
Crookes Collosols
  Company 200

Crookes electric lamp
  company 195
Crookes Healthcare 201
Crookes, Henry 192–201
Crookes Laboratories
  192, 201
Crookes, Madalina 195, 200
Crookes, William
  Chemical News 5, 73, 100,
    284, 286, 288
  "Chemistry of the
    Future" 42, 43–6
  chemistry reference
    books 247
  Crookes, Henry 193–7,
    199–200
  Dictionary of Chemistry 244
  Hofmann, August
    Wilhelm 284
  Huxley, T. H. 282–3
  "inmate of Hanwell
    asylum" 286–8
  picric acid 5
  poisonous dyes 6
  Prout, Edward 284
  Royal Society 200
  sons 193
  thallium 4–5
Crookes–Barnes
  Laboratories 192
Crum Brown, Alexander 104,
  106, 120, 154, 161, 165,
  262–3
Crum Brown–Frankland
  structural formulae
  notation 165
"cryohydrates" 107
Culinary Cookery 20
Cumming, James 279
Curtis, William 58

D'Albe, Fournier 200
Dalton, John 114–15, 117, 134,
  161, 163
Dalton's Law 147
Daniels, Farrington 265, 274
Darwin, Charles 60, 104
Das Ausland 77
Daubeny, Charles 80
Davidis, Henriette 22
Davies, Mansel 266
Davis, Jonas 239, 240
Davy, Humphry 25, 73,
  80, 158
Davy–Faraday Research
  Laboratories 14, 35–6
Dawes, Richard 58–60, 62, 63
De facultate sentiendi 9
de Kerlov, W. 52
de Lamétherie, Jean-Claude
  133
De Morgan, Augustus 24, 107
de Steiger, Isabelle 50–1
Death under Sail 312
Debye, Peter 274
Demarçay, Eugène 43
Dennis, Louis M. 271–3
Department of Science and Art
  (DSA) 63, 65–6, 111–12
deuterium oxide 228
Devonshire Commission on
  Scientific Instruction 110
Dewar, James
  Armstrong, Henry 35
  BAAS speech 43
  Crookes, Henry 200–1
  Jacksonian chair
    (Cambridge) 170
  Liveing, George 281
  low temperature physics 26
  pyridine 157

Royal Institution 33–9, 197, 199, 201, 281
  water quality 197
Dewar, Michael 183, 221
Dictionary of Applied Chemistry 247
Dictionary of Chemistry 100, 244–7
Dictionary of Chemistry (1795) 246
Dictionary of Chemistry and the Allied Branches... 246
Dictionary of Chemistry and Mineralogy 245
Dictionary of Scientific Biography 191
Die Aufgaben des geologischen Studium 76
Die Augsburger Allgemeine Zeitung 78, 79, 84
Dieffenbach, Ernst 74–81
Discovery 306
Ditmar, William 246
Dixon, Harold B. 181, 190, 264, 267
Djerassi, Carl 214
Donkin, William 162
Donnan, Frederick George 188–9
Drory, Stephanie 157
Druce, Gerard 53–4
Du Maurier, Daphne 289
Du Maurier, George 289
Du Val, Florence 215
Dulong and Petit's law 162–3
Dumas, Jean-Baptiste 115, 235–6
DuPont 41
Dye-Makers of Great Britain 7
"dynamite" 258
Eastwood, Eric 312

Eaton, Amos 86
Edgar, John George 282
Edison v. Swan patent case 196
Educational Times 110
Egerton, Alfred 298
Einfuhrung in die Mathematische Behandlung der Naturwissenschaften 263
Elements of Chemistry 288
Elements and Electrons 50
Elements of Experimental Chemistry 69
Elements of Inorganic Chemistry 244
Elements of Physical Manipulation 109
Elements of Physics 112
Eliot, Charles William 112
Eliot, Lewis (character) 313
Elliotson, John 12
Encyclopaedia of Chemical Technology 247
Engelmann, Wilhelm 234
Eno, James Crossley 298
eponymous chemical journals 233–6
Erlenmeyer 233
Escoffier, Auguste 15
ethylene (electrolytic oxidation) 221
Euclid 64
"Eureka" (Archimedes) 124
European Food Safety Authority 8
European Journal of Chemical Physics and Physical Chemistry (ECCP) 235
European Journal of Inorganic Chemistry (EURJOC) 235

European Journal of Organic Chemistry (EURJIC) 234–5
European Union (EU)
  chemical publications 234–5
  E-numbers 8
  food additives 8
  safety of chemicals 261
Evening Post (New York) 31
Experiments, Models, Paper
  Tools 115

Facts on Fire Prevention 259
Familiar Letters on Chemistry 78
Faraday, Michael
  benzene 18
  Brodie Jr, Benjamin
    Collins 159, 161
  Chemical Manipulation 86
  electromagnetic centenary
    celebration 174
  laboratory 87
  von Liebig, 80
  philosophical chemists 73
  Royal Institution 25
  state of matter 44
  "transmutation" 50–1
  Wöhler, Friedrich 142
Faraday Society 183, 186,
  188, 308
Ferguson, John 50, 52, 54
Field, Frederick 7, 118
Field's Yellow 7
Findlay, J. J. 173
Fire Risks in Industry 260
Firemen's Trade Union 259
Fisher, Nicholas 134
Fisica de corpi ponderabili 135
Fleck, Ludwig 237
Fleitmann, Theodor 150

Fleming, Ambrose 101
Flory, Paul J. 274
Food Standards Agency, UK 8
Forbes, Edward 118, 120
formonitrile (HCN) 223–4
Fortnightly Review 42
Foster, George Carey 97–102,
  103–4, 106, 113, 246
Foster, Michael 35
Fourneau 309 (sleeping
  sickness drug) 170
Fownes, George 244–5
Fox, Maurice 7
Francis, Frances 240
Francis, Rachel 241
Francis, William 118–19,
  241, 246
Frankland, Edward
  career 60
  Chapman, Ernest 289
  chemistry teaching 213
  dynasty 217
  graphic formulae 154, 165
  Guthrie, Frederick 103–5
  Kekulé, August 168
  *Lecture Notes for Chemical
    Students* 243, 288
  science education 64–5, 112
  structural notation 289
  valency 18, 120
  water analysis 288–9
  water quality 197
Franklin, Rosalind 102, 213
Fray Bentos Corned Beef 22–3
freemasonry 49
Freie Hessiche Zeitung 76
Fresenius, Carl Remigius
  143, 233
*Fresenius Journal of Analytical
  Chemistry* 236

Fresenius's Zeitschrift 233
Freund, Ida 211–13
Friedel–Craft reactions 203
Fry, Elizabeth 243
fuchsine (aniline red) 6

Gage, Alfred 112
Gardiner, James H. 195
Gardner, John 77, 81
Gay-Lussac, Joseph 88, 133,
135, 139–40, 162
Geilcher Lamp Co 196
*Genesis and Construction of a
Scientific Fact* 237
Geneva Convention 41
Geological Observer 77
Gerhardt, Charles 150,
161, 246
Getliffe, Frank 308
Ghosh, Jnanendra
Chandra 186
Gibsone, Burford Waring
279–80
Giebert, Georg 22
Gilbert, W. S. 211
Gillespie, Ronald 203
Gladstone, John Hall 122
Glaser, Karl 154
Glaucus 253
glutamic acid 14–15
"glyptic formulae" 18
Gmelin, Leopold 245
gold extraction 196
Goode, George Brown 30, 32–3
Gortner, Ross 274
Gosse, Philip Henry 249, 251–3
Govett, Ernest 199, 199–200
Graham, Thomas 103, 116,
142, 235, 244–5
Grammar of Chemistry 243

graphon (graphite radical) 160
"grease chemist"
(Schmierchemiker) 151
Great Britain
chemists 80
geologists 80
Great Exhibition 57
*Grossbritannien: Chemisches
Laboratorium in London* 79,
82–4
Groves, Charles E. 247
Guthrie, Frederick 60, 97,
102–13, 118, 246

H_2O (water) 114
Haber, Fritz 41, 176–9
Haber–process 183
Hague Convention, 1899 177
Halperin, John 305
Hampson, William 35
Handbuch der Chemie 245
*Handbuch der Organischen
Chemie* 141
Handel 283
Harcourt, Vernon 169
Hardy, William Bate 309
Harp Alley Boys' School 241
Harris, William A. 255–8, 261
Harrison, Tom 179
*Harris's Technological
Dictionary of Insurance
Chemistry* 255–6
Harvard University (chemistry/
physics teaching) 112–13
HAZCHEM signs 260
He Knew He Was Right 176
Heath, Grace 220
Heilbron, Ian 309, 311
Henrici, Olaus 173
Henry, William 69

Henslow, John 58, 60–2, 63
Hermetic Order of the Golden
    Dawn 50
heurism 171
heuristic teaching 124, 219
Hewitt, John 183
*Higher Mathematics for
    Chemical Students* 182,
    191, 264
*Higher Mathematics for
    Students of Physics and
    Chemistry* 182, 264, 268
Hildebrand, W. E. 269
Hill, Archibald Vivian 272
Hille, Hermann 192
*Hints on an Improved System of
    National Education* 59
Hirst, Thomas Archer 100
Historical Record (University of
    London) 283
*History of Chemical Theory
    from the Age of Lavoisier* 247
History of Chemistry 191
HMS Tory 76
Hodgkin, Thomas 76–7
Hodgkins, Thomas George
    26–39, 73
Hodson, William 279
Hofmann, August Wilhelm
    ammonia structure 287–8
    "aniline violet" 6
    Armstrong, Henry 168
    biography 16–19
    Brodie Jr, Benjamin
        Collins 160
    Crookes, William 284
    Hofmann, Wilhelm (von
        Liebig's pupil) 88, 90
    *Introduction to Modern
        Chemistry* 17–18

Kekulé, August 151, 155–6
    marriage 105
    Royal College of Chemistry
        118, 143
    Royal Institution 18, 90, 174
Hofmann, Paul 88, 92–3, 95–6
Högel, Luise 157
Holland, Kathleen 214–15
Holland, Lily 214–15
Holland, Mina 214–15
Holland, William 215
Holloway, Thomas 220
Holmyard, Eric 106
Home, Daniel Dunglas 193
homologous term 99
Hooke's law 108
Hopkins, Gowland 302, 309
*Hoppe-Seyler's Zeitschrift für
    physiologische Chemie* 236
Horlicks 50
How to Teach Chemistry 213
Hughes, Edward 140, 228
Hunter, John 77
Hunter, Louis 302–4
Huxley, T. H. 60, 103, 105, 168,
    170, 213, 282–3
hydrogen cyanide
    composition 223
    specific heats 187, 222, 224
    Zyklon process 177
hydrogen sulfide 90

Ikeda, Kikunae 13–15
Imperial College of Science &
    Technology 169, 172–3
In Memorium (poem) 46
Industrial Ceramics 270
*Industrial and Engineering
    Chemistry* 310
Industrial Revolution 74

Ingle, Herbert 258
Ingold, Christopher Kelk
  Armstrong, Henry 167
  biography 187
  Hughes Edward 140
  Partington, J. R. 188
  Structure and Mechanism in
    Organic Chemistry 274
  Usherwood, Edith
    Hilda 187, 218–30
  Wilson Christopher 203
Ingold, Dilys 228
Ingold, Keith Usherwood 228
Ingold, Sylvia 228
"inmate of Hanwell
  asylum" 286–9
Institute of Structural
  Engineers 259
insurance chemistry 255–61
International Academy of
  Mathematical Chemistry 266
Introduction à l'Etude de Chimie
  par la Système Unitaire 150
Introduction to Modern
  Chemistry 17
Investigation of the Laws of
  Thought 162
Ionides, A. C. 194
isologous term 99
isomerism 139, 222

Jahresberichte 143–4
Jamin, Jules 99–100
Japp, Francis 154
"jargonthropos" 167
"Jellite" 224
Jevons, William Stanley 57–8
Johnston, J. F. 242
joint research papers 140
Jollivet-Castelot, François 52

Jones, Marian 183
Jones, William 69, 71–2
Journal (Alchemical Society) 53
Journal of Chemical
  Education 174
Journal of the Chemical
  Society 223, 262
Journal (London Chemical
  Society) 245
Journal of Mathematical
  Chemistry 266
Journal de Physique 133
Journal of Physical
  Chemistry 174, 186
Journal für Praktische
  Chemie 233
Journal of the Royal Society of
  Chemistry 236
journals (chemistry) 233–6
Jowett, Benjamin 161
Julius (Duke of Braunschweig-
  Wolfenbüttel) 209–10

Kabbalah (religion) 54
Kane, Robert John 80
Katz, Johann Rudolf 273
Keats, John 242
Kekulé, August
  architecture to
    chemistry 148–50
  Armstrong, Henry 168
  Bonn 156–7
  chemistry of radicals 141
  family life 157
  Foster, George Carey 99
  Ghent 154–6
  introduction 147–8
  London 150–2
  organic chemistry 139, 145
  types to catenation 152–4

King, Frederick E. 190
Kingsford, Anna 49
Kingsley, Charles 253
Kipping, Frederic Stanley 215–17
Kipping, Lily 217
Kipping, Stanley 43
Kipp's apparatus 94
Kirk, Raymond Eller 247
Klein, Ursula 115
Knapp, Friedrich 78, 247
Knight, John 68
Knorr, Ludwig 178
Koch, Robert 96
Kolbe, Hermann 90–4, 103,
    141, 156, 168–9, 233
Kopp, Hermann 40–1
Körner, Wilhelm 154, 157
Kraftküche von Liebigs
    Fleishextract 22
Krahe, Eduard 214, 217
Kritische Zeitschrift für
    Chemie 154
Krüss, Gerhard 233
Krystallsysteme und
    Krystallstructur 263
Kuhn, T. S. 189
Kunckel, Johann 202
Kynch, G. J. 266

La Grande Illusion 179
La Guide Culinaire 15
La Société Alchimique de
    France 52
Laar, Peter 222–3
laboratories
    "Cathedrals of Science" 96
    chemistry's influence on
        other sciences 95–6
    von Liebig 85–90
    post von Liebig 90–4

Laboratory 97, 147–8
Lacey, Henry 68
Ladenburg, August 156
Laidler, Keith 180
Langley, Samuel P. 26, 30,
    37, 38
Laplace, Simon 134, 140
Laputa (imaginary
    country) 164
Lapworth, Arthur 215–17, 226
Lathbury, Kathleen 226
Laurent, August 151, 153
Lavoisier, Antoine 85, 132,
    140, 161
Le Fèvre, J. W. 190
Leadbeater, Charles 44
Lecture Notes for Chemical
    Students 168, 243, 288
Lectures on History 241
Lectures on Morbid Anatomy of
    Serous and Mucous
    Membranes 76
Leffek, Kenneth 187
Lehrbuch der Chemie 143
Lehrbuch der organischen
    Chemie 154
Les Nouveaux horizons de la
    science et de la pensée 52–3
Lewis, Gilbert Newton 202,
    225, 273
von Liebig, Justus
    An Address to Agriculturalists
        of Great Britain...Artificial
        Manures 79
    Animal Chemistry 115
    Annalen der Chemie... 233,
        235–6
    condenser 94, 208
    Dieffenbach, Ernst 75, 77–81
    food industry 20–3

Francis, William 246
Gardner, John 77
Hofmann, A. W. 16
Kekulé, August 150–2
laboratories 85–7, 87–90
"Liebig on Toast" 23
organic analysis 115
Wöhler, Friedrich 138–46
"Liebig sandwiches" 23
"Liebig on Toast" 23
Liebigs Annalen 236
Linnaean Society 240
Lister, Joseph 168
Liveing, George Downing
279–81
Lockyer, Norman 43, 108
Lodge, David 289
Lodge, Oliver 101–2, 103
Logrono 105
London Chemical
Society 67–73, 245
London Fire Brigade 260
*London Medical and Physical
Journal* 12
Lonsdsale, Kathleen 229
Loschmidt, Josef 136–7
Loschmidt's number 137
Low Temperature Research
Fund 34
Lowell, James Russell 272
Lowry, Martin 202, 227, 230,
306, 308
Luggin, Hans 178
Lummer, Otto 182–3
"Luzitanus" 69
Lyell, Charles 77

MacArthur, J. S. 196
"magic acid" 203
Magnus, Gustav 97

"Manchester Yellow" 5, 7
Manual of Chemistry 244
Maoris (New Zealand) 77
Marcet, Alexander 214
Margenau, Henry 266
"Marquess of Hastings"
colours 4
Marreco, A. J. F. 70–1
Martius, Carl 5
Mary the Jewess 207–8
Maskelyne, Nevil Story 159
mass spectroscopy 141
Massachusetts Institute of
Technology (MIT) 109–10
math for chemist 262–6
*Mathematical Methods for
Science Students* 266
*Mathematical Preparations for
Physical Chemistry* 265
*Mathematics of Physics and
Chemistry* 266
Mathias, E. 297
Mathiessen, Augustus 99, 170
"Mathilda effect" 218
*Matter, Spirit and the
Cosmos* 53
"Matthew effect" 218, 230
Maxwell, James Clerk
97–8, 108
McVeigh, Dr 4, 6
Mechanics' Institute 67–8
Mechanics' Magazine 68, 73
Medusae 249
Meldrum, Andrew
Norman 181
Mellor, Joseph William 182,
189, 263–6, 267–70
Mendeleev Communications 236
Menzies, A. C. 302–5
Merchant Venturers College 59

mercury 207–8
mercury fulminate 152
Messiah 283
"methyllic nitrate" (ethyl nitrate) 289
Meyer, Victor 165
Miall, Stephen 175
Michael condensation 227
Miles, Arthur (character) 312
Miller, Frederick B. 193
Miller, William Allen 193, 288
Mills, W. H. 306, 307
Modern Chemistry 120
Modern Cookery 21
Modern Opera Houses and Theatres 259
Moldenhauer, Friedrich 149
Mond, Ludwig 36
Mongredien, August 72
monosodium glutamate (MSG) 15
Moore, Bernard 268
Moore, S. 141
Moore, Thomas S. 67–8, 221
Morley, John 4, 7
Morris, Malcolm 201
Morton, Richard 309, 311
Moseley, H. G. J. 59
Moseley, Henry 59
Moseley, Henry Nottidge 59
motion and gustation 12
Müller, Hugo 151
Mulliken, Robert S. 307–8
Murphy, G. M. 266
Muscular Movements in Man 272
musical affinities 214–17
Muspratt, James 21, 78–9
Muspratt, James Sheridan 79
Muspratt, Richard 79

naphthalene dyestuffs 170
Native Guano Company 193–4
Nature
  Alchemical Society 52
  Armstrong, Henry 168
  future of chemistry 42–3, 310, 312
  Guthrie, Frederick 105, 107, 110–11
  Ingold, Hilda
    deuterium oxide 228
    dinitrogen tetroxide 229
  Nernst's heat theorem 183
  Ramsay, William and radon 50
  Snow, C.P.
    photochemistry of vitamins 310
    UV absorption in aldehydes 312
Needham, Joseph 185, 312
Neoplatonism 49
Nernst, Walther 41, 182–3, 186, 191, 234, 263–4, 266
Neufeldt, Sieghard 43
New Lives for Old 312
Newcastle Chemical Society 71
Newnes, George 15
Nicholson, Edward 18
Nicholson, William 246–7
Nicholson's Journal 233
nitroglycerine 5, 257
nitrophenylenediamine (red dye) 6
NMR 141
Nobel, Alfred 257–8
nuclear magnetic resonance (NMR) 204
Nuffield School Science Project 125

Nummedal, Tara 210
Nyholm, Ronald 229

"occult chemistry" 44–5
Occult Club, Piccadilly 52
Odling, William 106, 120–1, 151, 156, 196–7, 246, 295
Olah, George 203–4
Old, William Gorn 51
"old wood dyes" 3
Oldenburg-Verlag (publishers) 235
olfaction 11
On Furnaces and Apparatus 207
On a Gold Basis 50
On the Utilization of Sewage by Phosphate of Alumina 120
Once a Week 292
Organic and Bioorganic Chemistry - A European Journal 234–5
ornithopachynispaedeia (cramming of children) 63
Ostwald dilution law 182
Ostwald, Wilhelm 14–15, 168, 198, 233–5, 263
Othmer, Donald 247
Outline of Experiments and Description of Apparatus and Material... 112
Owen, Alex 49
Owen, Richard 76
Owen, Robert 60
Oxford Dictionary of National Biography 289
Oxo 23

"paper chemistry" 115
Paradise Lost 282
Paris Chemical Society 155

Paris Exhibition, 1867 74, 107
Parker, Leslie Henry 183
Parkes, Samuel 214, 242–3
Parthasarathy, R. 186
Partington, Charles Frederick 71, 73
Partington, James Riddick
    conclusions 190–1
    early career 180–6
    Faraday Society 183, 186
    introduction 180
    publications 264–6
    research in physical chemistry 186–8
    Royal Society 188–90
    specific heat on hydrogen 226
Pasteur, Louis 85, 96, 168
Patterson, Hubert Sutton 296, 298
Pauling, Linus 47, 273–4
Peel, Robert 78
Peene, William Gurden 24
Pennington, Isaac 279
Pepys, Samuel 125
Pepys, W. H. 88
Perkin, Mina 217
Perkin, William Henry 43, 122, 215
Perkin Jr, William Henry 181, 267
Perrin, Jean 137
Perry, John 95, 103, 106, 110, 170, 173
Peter Ibbotson 292
Pharmaceutical Journal 82
Pharmaceutical Society 90–1
Pharmaceutical Times 82
Philip, J.C. 67–8
Phillips, Richard 243

Phillips, William 73, 80
Philosophical Magazine 73, 118, 226, 238, 240, 246
Philosophical Transactions 114
Phipson, Thomas Lamb 279
Phoenix Assurance Company 255
Phosphorescence 280
Physical Chemistry 191
Physical Laboratory Practice 113
Physical Manipulation 113
Physico-Chemical Metamorphoses... Piezochemistry 272
Physikale Zeitscrift 183
Physiology 12
Pickering, Edward Charles 95, 109–10, 112–13
picric acid (aniline yellow) 4–5, 7
Pinoh's Restaurant, London 52
Playfair, Lyon 63, 104
Plimmer, Henry 200
Poggendorff's Annalen 142
poisonous socks 3–8
Pope, William Jackson 43, 188, 201, 281, 306, 307
Potter, Benjamin 100
Potter, Theresa 294
pottery chemistry 267–70
Practical Cookery Book 22
Practical Physics, Molecular Physics and Sound 108, 112–13
Priestley, Joseph 241, 248
Prince Albert 65, 78
Princess Ida 211
Principles of Polymer Chemistry 274–5

Pringsheim, Ernst 182–3
Proceedings (Royal Institution) 35
Proceedings (Royal Society) 312
Proust, Louis-Joseph 135
Prout, Ebenezer 283–5
Prout, Edward Stallybrass 284–5
Prout, William 9–13, 15, 44, 73, 139, 285
Psychological Enquiries 158
Punch 289, 293
pyridine 156–7

Quarterly Journal of Science 5, 42
Queen Victoria 78
Quincke, Georg 99–100

radioactive disintegration 44–6
radium 197–8
Rainbow Tavern, Fleet Street 118
Ramsay, William
  Armstrong, Henry 167
  Chemical Society 42
  Cripps, Stafford 295, 299
  death 54
  Dewar, James 35
  future of chemistry 43–7
  Lord Rayleigh 36–7
  Occult Club 52
  radon 46, 50
  Royal Society 297
  transmutation 46
  xenon 296–7
Random House 52
Ratthausen, Heinrich 14
Rayleigh, Lord 36–7, 42, 45, 47

REACH (*R*egistration, *E*valuation, *A*uthorization and Restriction of *Ch*emical Substances) 260–1
Reckitt Benckiser 200–1
"Recollections of an English goldmine" 292
red dyes 5–6
Redgrove, Herbert Stanley 48–9, 52–4
Regnault, Henri 97
Reid, David Boswell 58, 60–2, 86
Reimann, Max 4–5
Remsen, Ira 143
Reports of the British Association for the Advancement of Science 238
Reports of Progress in Applied Chemistry 311
Researches on the Hydrides of Boron and Silicon 274
Richardson, Thomas 247
Richter, Jeremiah Benjamin 262
Rideal, Eric K. 183, 188–9, 306, 307, 309
Robertson, Mary 221
Robertson, Robert 308
Robins, E. C. 170
Robins, Edward Cookworthy 59, 65
Robinson, Gertrude 218
Robinson, Robert 167, 218, 226
Rochow, Eugene 214, 217
Rocke, Alan 151, 153, 168
Rogers, William Barton 109
Roscoe, Henry 90, 97, 103, 170, 181

Royal Chemical Society 67
Royal College of Chemistry (RCC) 16–18, 74, 78, 80–2, 118, 143, 168, 284, 289
Royal College of Physicians 24
Royal Institution
  Armstrong, Henry 35–6, 174
  Brodie Jr, Benjamin Collins 159
  "Charcoal Period" 36
  Crookes, Henry 201
  Crookes, William 199, 201
  Davy–Faraday Research Laboratories 14, 36
  Dewar, James 33–9, 197, 199, 201, 281
  Hodgkin's Trust 32–36
  Hofmann, August Wilhelm 18, 90, 174
  Ikeda, Kikunae 14
  von Liebig, Justus 20
  research 24–39
  Smithsonian Institution 30
  ventilation of laboratories 90
Royal Society
  Animal Chemistry Club 67
  Baker lectures 271–5
  Calculus of Chemical Operations 162
  Cripps, Stafford 300
  Crookes, Henry 197
  Crookes, William 200
  Hodgkins, Thomas George 39
  Partington, J. R. 180–90
  Pepys, Samuel 125
  Philosophical Transactions 114
  Proceedings 312
  Ramsay, William 297

Royal Society (*continued*)
  science education 63
  Snow, Charles Percy 311
Royal Society of Chemistry
  journals 235–6
  Warington, Robert 248
  women members 71
Rudorf, George 296
Russell, Colin 157
Russell, William James 118–19,
  246
Russian Academy of
  Sciences 236
Rutherford, Ernest 43, 45–6,
  50, 302
Rutt, J. T. 241

'S' and 'three R's'
  introduction 57–8
  science
    elementary teaching 58–63
    secondary schools 63–6
Sachs, Edwin Otto 258–9
Sakurai, Jōji 14
salderos (beef salting plant) 22
Sanderson, F. W. 219
Sandhurst military academy 65
Sayers, Dorothy L. 307
Schlorlemmer, Carl 153
Schoenflies, Artur Moritz 263, 266
Schombach, Heinrich 208
Schweitzer (mineral water
  manufacture) 251–2
Science Citation Index 234
"scientific education" 57
  see also 'S' and 'three R's'
Secondary Education Act,
  1902 62
*Selected Topics in Colloid
  Chemistry* 274

Shaw, Florence 227
Shenstone, William 106
Sheppard, Harry 53
Shilling, W. G. 187, 226
Shirley, Ralph 50, 53
Shuttleworth, Kay 120
Sidgwick, Nevil 188, 274
"Sign of the Hexagon" 174
silver collosols ("Silvagen") 201
silver cyanate 140
Singer, Felix Gustav 270
Singer, Sonja 270
smell loss (anosmia) 11
Smith, George 72
Smith, Goldwin 272
Smith, Henry 162
Smith, James Edward 240
Smith, Miles H. 288
Smithson, James 37, 38
Smithsonian Institute,
  Washington 24–9, 30–39
Snow, Charles Percy
  Cambridge 305–6
  introduction 301–4
  novelist 312–13
  photochemsitry of
    vitamins 310–11
  physical chemistry 306–10
  provinces to mecca of British
    science 302–5
  Second Law of
    Thermodynamics 131
  why science? 305
Society for Study of Alchemy
  and Early Chemistry 48
socks, poisonous 3–8
Soddy, Frederick 43, 45, 50–2
*Some Applications of Organic
  Chemistry to Biology and
  Medicine* 273

Some Physical Properties of the Covalent Link in Chemistry 274
Sömmering, Philp 209
Spagnoletti, C. E. 195
Specific Heats of Gases 191
Spencer, Herbert 57–8, 66, 158
Square Rounds 179
Stallybrass, Edward 283
Stanley, Arthur 161
Stas, Jean-Servais 42, 154
Steenbock, Henry 309
Stein, W. H. 141
Stenhouse, John 150
Stephenson, Geoffrey 266
stereoisomerism 18
Sterne, Lawrence 107
Stewart, Alfred, W. 214
"stinks" (smells) 90
Stock, Alfred 274
Strangers and Brothers 308, 312
Strauss, Ludwig 211
Strecker, Adolphe 150
Stroud, Leslie 199–200
"structural models" 18
Structure and Mechanism in Organic Chemistry 274
substitution concept 151
Sudhoffs Archiv, Zeitschrift für Wissenschaften 236
Suggestive Inquiry into the Hermetic Mystery 50
"superacids" 203
Svedberg, The 198
Swedenborg, Emmanuel 53
sweet, sour, bitter and salty tastes 10
Swiney, George 25
Swithinbank, Isobel 298–9

Tales of Hofmann 17
tartrazine 7–8
taste, smell and flavour 9–15
tautomerism 223, 225, 227, 230
Taylor & Francis 238–9
Taylor & Hessey 242
Taylor & Walton 242
Taylor, A. M. 308
Taylor, Harold 273
Taylor, John 239, 242
Taylor, Richard 238–43
Taylor, Susannah 239
teaching chemistry to women 211–13
Technical School and College Buildings 65
Technologie 247
Temple, Frederick 64
Tennyson (poet) 46
textbooks of chemistry 237–43
The Aquarium 252–4
The Blue Angel 179
The Boyhood of Great Men 282
The Chemical Catechism 243
The Chemist 68, 70–3
The Chemistry of Food 20–1
The Cream of Beauty 54
The Devoted 305
The Documents in the Case 307
The Double Helix 102
The Experimental Basis of Chemistry 213
'The Future Allchemist' 174–5
The Gypsies of Spain 105
The Hindu 186
The Holland Sisters 214, 217
The Jew 105
The Lancet 4, 6, 82, 310
The Manhood of Great Boys 282

The Martian 293
The Occult Review 50, 53, 54
The Origins of Alchemy 51
*The Place of Enchantment:
    British Occulism and the
    Culture of the Modern* 49
*The Rudiments of
    Chemistry* 243
The Search 301, 306, 312
The Specific Heats of Gases 187
*The Study of Chemical
    Composition* 212–13
*The Teaching of Science at
    Christ's Hospital since 1900
    A.D.* 124
The Times 3, 5–6, 168
"the vivarium" 250
*The Wave Mechanics of Free
    Electrons* 273
The Westminster Gazette 52
The World Set Free 50
Thenard 88, 140
Theoretische Chemie 183
Theosophical Society 44, 49
Thermodynamics 183, 191
Thomas, Mary Beatrice 213
Thomson, George P. 273
Thomson, J. J. 37
Thomson, Thomas 80
Thomson, William 97–8,
    249–51
*Thomson's Annals of
    Philosophy* 233
Thorp, William 247
Thorpe, Jocelyn Field 222,
    224, 227
Thorpe, Thomas Edward 40,
    170, 173, 247
Threlfall, Richard E. 294
Thudichum, John 138

Thynne, Anna 251
Tidy, Charles 196
Tilden, William 170
Tilloch, Alexander 238, 240
*Tilloch's Philosophical
    Magazine* 233
*Transactions of the Chemical
    Society* 181
*Transactions of the Faraday
    Society* 228
Transactions (Linnaean
    Society) 240
*Transactions of the Newcomen
    Society* 236
Trautschold, Wilhelm 80
Travels in New Zealand 75
Travers, Morris W. 37, 47, 296
*Treatise on the Ceramics
    Industry* 269
Treatise on Thermodynamics
    264
Trilby 292
Tristram Shandy 107
Trollope, Anthony 176
Tswett, Mikhail 43
Turba Philosophorum 51
Tyndall, John 35, 60,
    105–6, 168

*Übung in der analytischen
    Chemie* 143
umani ("delicious") taste
    13–15
*Uncle Joe's Nonsense for Young
    and Old* 270
University of Cambridge
    appointments 279
uranium 44–5
Ure, Andrew 245, 247
Urey, Harold 273

Usherwood, Edith Hilda
Ingold partnership 218–30
Partington, J. R. 187
thesis 222
Usherwood, Thomas S. 219

"valency" concept 151, 157
van Laar, Conrad 223
van Praagh, Gordon 124–5
van Voorst, John 243
van't Hoff, J. J. 14, 145, 154,
156, 164, 168, 234, 272
vapour density 133
Vernon Harcourt, A. G. 221
Victoria Gold and Copper
Mine 292
"Victoria Orange" 5
Vieweg, Edouard 77
Vogel, Arthur Israel 190
Volta, Alessandro 132
von Nudeln, A.
(pseudonym) 107
von Oettingen, Carl
(fictitous) 209
von Planta, Adolf 150
von Ringen, Hugo 80
von Stroheim, Erich 179

Wagner Meerwein
rearrangements 203
Waite, Arthur 50–1
Wakley, Thomas 81
Wanklyn, Alfred J. 288–9
Ward, Nathaniel Bagshaw
248, 251
"Wardian case" 248
Warington, Robert 72–3,
248–54
Warington Jr, Robert 68
Wasserman, Albert 228

water bath (*bain marie*) 207–8
water purity (London) 197
Watson, James 102
Watts, Henry 100, 103, 244–7
Webber, Dr 3–4, 6
Wellby, Philip 50–2
Wells, H. G. 42, 50, 102, 106,
110, 113
Weltzien, Karl 154
West Indian Volcanoes
Committee 197
Whewell, William 59, 117
Whitaker, H. 228
White, Andrew D. 272
Whiteley, Martha 221–3
Whiteley, Martin 187
Whytlaw-Gray, Robert 228,
296, 298
Wiley (publishers) 235
Wilks, Richard 240
Will, Heinrich 142–3, 150
William IV (King) 242
William Oppenheimer
(manufacturing
chemists) 200
William Rider & Co. 51–3
Williamson, Alexander 99–100,
103, 151, 153, 156, 160–1, 165,
245, 288, 291
Willstätter, Richard 41, 43
Wilson, Christopher 203, 228
Wilson, Harold 313
Wilson, James 61–4, 66
Windler, S. C. H. 286
Winstein, Saul 203–4
Wöhler, Friedrich 115–16,
138–46, 286
Wollaston, William Hyde
73, 80
Wolseley, John 79

women
    alchemy and chemistry 207–10
    teaching chemistry 211–13
Woolwich military academy 65
*World of Physical
    Chemistry* 180
Worthington, Arthur 106, 113
Wurtz, Adolphe 86, 152–3, 161,
    194–5
Wurtz, Charles 247
Wynne, W. P. 170

yellow dyes 4–5, 7
Yoshioka, Kannosuke 112
Youmans, William Jay 32

Young, Hannah 22–3
Young, Sydney 40–1, 297
Youth Searching 305

*Zeitschrift für analytischen
    Chemie* 233
Zeitschrift für Chemie 233
*Zeitschrift für physikalische
    Chemie* 234–5
Zieglerin, Anna Maria
    208–10
Zonite Products
    Corporation 192
Zyklon process (hydrogen
    cyanide) 176